Springer Series in
Surface Sciences

12

Editor: Gerhard Ertl

Springer Series in **Surface Sciences**

Editors: Gerhard Ertl and Robert Gomer

Volume 1: **Physisorption Kinetics**
By H. J. Kreuzer, Z. W. Gortel

Volume 2: **The Structure of Surfaces**
Editors: M. A. Van Hove, S. Y. Tong

Volume 3: **Dynamical Phenomena at Surfaces, Interfaces and Superlattices**
Editors: F. Nizzoli, K.-H. Rieder, R. F. Willis

Volume 4: **Desorption Induced by Electronic Transitions, DIET II**
Editors: W. Brenig, D. Menzel

Volume 5: **Chemistry and Physics of Solid Surfaces VI**
Editors: R. Vanselow, R. Howe

Volume 6: **Low-Energy Electron Diffraction**
Experiment, Theory and Surface Structure Determination
By M. A. Van Hove, W. H. Weinberg, C.-M. Chan

Volume 7: **Electronic Phenomena in Adsorption and Catalysis**
By V. F. Kiselev, O. V. Krylov

Volume 8: **Kinetics of Interface Reactions**
Editors: M. Grunze, H. J. Kreuzer

Volume 9: **Adsorption and Catalysis in Transition Metals and their Oxides III**
By O. V. Krylov, V. F. Kiselev

Volume 10: **Chemistry and Physics of Solid Surfaces VII**
Editors: R. Vanselow, R. Howe

Volume 11: **The Structure of Surfaces II**
Editors: J. F. van der Veen, M. A. Van Hove

Volume 12: **Diffusion at Interfaces: Microscopic Concepts**
Editors: M. Grunze, H. J. Kreuzer, J. J. Weimer

M. Grunze
H. J. Kreuzer J. J. Weimer (Eds.)

Diffusion at Interfaces: Microscopic Concepts

Proceedings of a Workshop
Campobello Island, Canada, August 18–22, 1987

With 126 Figures

Springer-Verlag Berlin Heidelberg New York
London Paris Tokyo

Professor Dr. Michael Grunze
Dr. Jeffrey J. Weimer

Laboratory for Surface Science & Technology, Department of Physics,
University of Maine at Orono, Orono, ME 04469, USA

Professor Dr. Hans Jürgen Kreuzer

Department of Physics, Dalhousie University, Halifax, Nova Scotia, Canada, B3H3J5

Series Editors

Professor Dr. Gerhard Ertl

Fritz-Haber-Institut der Max-Planck-Gesellschaft, Faradayweg 4–6
D-1000 Berlin 33

Professor Robert Gomer

The James Franck Institute, The University of Chicago, 5640 Ellis Avenue,
Chicago, IL 60637, USA

ISBN-13: 978-3-642-73634-6 e-ISBN-13: 978-3-642-73632-2
DOI:10.1007/978-3-642-73632-2

Softcover reprint of the hardcover 1st edition 1988

Printing Druckhaus Beltz, 6944 Hemsbach/Bergstr
Binding J Schaffer GmbH & Co KG , 6718 Grunstadt
2153/3150-543210

Preface

The second workshop on Interface Phenomena was organized jointly by the surface science groups at Dalhousie University and at the University of Maine. Continuing the philosophy of the previous workshop on the Kinetics of Interface Reactions, published as Volume 8 of the Springer Series in Surface Sciences, we concentrated on a few selected topics related to Diffusion at Interfaces.

Diffusion in bulk materials and at interfaces has been studied at the macroscopic level for over a century; the theoretical framework was formulated by Fick in 1855 when he wrote down the diffusion equation. An understanding of diffusion phenomena at the microscopic level emerged at the turn of the century with Einstein's theory of Brownian motion, and subsequently from Perrin's experiments. Diffusion coefficients in dilute gas mixtures were eventually calculated from Boltzmann's equation after the perfection of the Chapman-Enskog method. In contrast, diffusion in solids and at interfaces remains a very active area of research as the contributions in this volume attest.

The workshop was supported by the Air Force Office of Scientific Research, The Office of Naval Research, the Surface Science Division of the Canadian Association of Physicists, and the University of Maine. The views, opinions, and/or findings contained in this report are those of the authors and should not be construed as an official position, policy, or decision by any of the above agencies, unless so designated by other documentation.

We would also like to thank the following companies for their donations to help sponsor graduate student attendance:

GTE Products Corporation
Howell Laboratories
MDC Vacuum Products
SPECS GmbH
Standard Oil Corporation of Ohio
Varian Associates

The workshop was again held at the Roosevelt Campobello International Park on Campobello Island, New Brunswick. The superb accommodation in the various "cottages" within the Park and the serene beauty of the Bay of Fundy provided an ideal setting for intensive discussions. Speaking on behalf of all participants, we would like to express our gratitude to the staff of the Park for making our stay there so pleasant.

Orono and
Halifax,
January 1988

M. Grunze
H.J. Kreuzer
J.J. Weimer

Contents

Part I Diffusion at the Vacuum-Solid and Gas-Solid Interface

Surface Diffusion Measured Using Laser Induced Thermal Desorption: Hydrogen on Ru(001)
By S.M. George (With 13 Figures) 2

Reflection High Energy Electron Diffraction Studies of Diffusion and Cluster Formation During Molecular Beam Epitaxy
By P.R. Pukite, G.S. Petrich, G.J. Whaley, and P.I. Cohen (With 15 Figures) 19

Laser-Induced Desorption Determinations of Surface Diffusion on Rh(111)
By E.G. Seebauer and L.D. Schmidt (With 3 Figures) 37

Mobility of Rare Gases Adsorbed on the (001) Surface of MgO
By M. Karimi and G. Vidali (With 2 Figures) 43

Comparison of Equilibrium and Non-Equilibrium Diffusion Measurements for W(110)p(2×1)-O
By M.C. Tringides, P.K. Wu, and M.G. Lagally (With 5 Figures) 53

Hopping and Diffusion
By H.J. Kreuzer 63

Diffusion at the Vacuum-Solid and Gas-Solid Interface: Some Comments
By J.H. Block 67

Part II Diffusion at the Solid-Solid Interface, Surface Segregation and Bulk Diffusion

Diffusion at Surfaces: Introductory Remarks
By E. Zeitler 72

Direct Observation of Atomic Motion on Surfaces
By H.-W. Fink (With 13 Figures) 75

Topography Modification and Microscopic Motion on Metal Surfaces
By R.J. Behm (With 7 Figures) 92

Surface Self-Diffusion, Capillarity, and Surface Steps
By J.M. Blakely and C.C. Umbach (With 5 Figures) 102

Radiation Enhanced Diffusion (RED) in a Sputtered Ag/Ni Layered System
By D. Marton, J. Fine, and G.P. Chambers (With 3 Figures) . . . 111

The Influence of Segregation on the Formation of Cu/Ni Alloy Films by Means of Evaporation
By B. Hofmann, D. Borgmann, and G. Wedler (With 10 Figures) 121

Surface Properites of a Lead-Tin Alloy
By W.N. Unertl, M. Grunze, and J.J. Weimer (With 7 Figures) . 132

Part III **Theoretical Models for Segregation and Diffusion**

Recent Developments in the Theory of Many-Body Processes: Implications for the Study of Surface Dynamics
By J.D. Doll, D.L. Freeman, and A.F. Voter (With 1 Figure) . . . 144

Computer Simulation of Segregation and Diffusion at Oxide Interfaces
By P.W. Tasker and W.C. Mackrodt (With 6 Figures) 155

Part IV **Nucleation at the Solid-Liquid Interface**

Geometrical Structure of Adlayers on Immersed Electrodes
By J.G. Gordon II, O.R. Melroy, and L. Blum (With 8 Figures) . 172

Electrochemical Processes in Small Systems
By M. Fleischmann (With 28 Figures) 182

Index of Contributors 205

Part I

Diffusion at the Vacuum-Solid and Gas-Solid Interface

Surface Diffusion Measured Using Laser Induced Thermal Desorption: Hydrogen on Ru(001)

S.M. George

Department of Chemistry, Stanford University,
Stanford, CA 94305, USA

The surface diffusion of hydrogen atoms was studied on clean and chemically-modified Ru(001) surfaces using laser induced thermal desorption techniques. Experiments at low coverages between 260-330 K revealed an activation barrier for surface diffusion of 4 kcal/mole and a preexponential of 6.3×10^{-4} cm ^{2}/sec. The diffusion coefficients were not coverage-dependent over a wide range of coverages. The isotope effect was close to the predictions from simple transition state theory suggesting that quantum mechanical tunneling effects are minor in the measured temperature range between 260-300 K. Carbon and sulfur on Ru(001) were both found to reduce the surface mobility of hydrogen atoms. The effect of carbon was explained by a trapping mechanism, whereas the effect of sulfur was consistent with a blocking mechanism. Coadsorbed carbon monoxide also reduced the surface mobility of hydrogen and displayed a pronounced reduction at a coverage where an ordering transition occurs in the CO adlayer.

I. Introduction

Surface diffusion plays a central role in many surface processes. [1]. Surface diffusion studies date back to the 1920's when Volmer and Estermann [2] noticed a pronounced anisotropy in the growth of mercury crystals. Since these early observations, numerous experimental methods such as field ion microscopy (FIM) [3] and field emission microscopy (FEM) [4] have been employed to examine surface diffusion of atomic and molecular adsorbates on field emitter tips. With the advent of modern UHV techniques on macroscopic single-crystal surfaces [5], coverage gradient surface diffusion experiments utilizing scanning Auger [6] or work function probes [7] have been conducted but have been restricted to tightly bound atomic adsorbates.

Surface diffusion experiments in UHV on macroscopic, well-defined single-crystal surfaces are desirable because the surface order and composition can be verified and controlled. Consequently, the effect of surface order and composition on surface diffusion can be examined. These surface diffusion experiments are both theoretically and technologically important. For the theory of surfaces, surface diffusion reveals the underlying adsorbate-surface and adsorbate-adsorbate interactions that define the surface potential. For heterogeneous catalysis on surfaces, surface diffusion as a function of surface cleanliness or adlayer composition may effect surface reaction kinetics.

Until recently, the surface diffusion of various atomic and molecular adsorbates could not be measured on macroscopic, single-crystal surfaces. Since the development of laser-induced thermal desorption (LITD) techniques [8-13], this situation has changed. Now the surface diffusion of various chemically important adsorbates can be measured on well-characterized, macroscopic surfaces. Moreover, because the LITD technique commonly employs a mass spectrometric detection system, the identity of the diffusing species can be clearly established.

Following a description of the surface diffusion experiments using LITD, this paper will focus on the surface diffusion of hydrogen on clean and chemically-modified Ru(001) surfaces. The Ru(001) surface is a model, close-packed transition metal surface and has proven catalytic abilities, especially for methanation and Fischer-Tropsch synthesis [14,15]. Likewise, hydrogen is present on transition metal surfaces in various hydrogenation and dehydrogenation catalytic steps and is considered a prototypical adsorbate. Consequently, these hydrogen surface diffusion studies on clean and chemically-modified Ru(001) surfaces should be interesting to both surface theorists and surface kineticists.

II. Surface Diffusion Measured Using Laser-Induced Thermal Desorption

Classical diffusion experiments measure the time-dependent relaxation of a concentration gradient and employ a Fick's Law analysis in order to obtain the diffusion coefficient [16]. The well-known lecture demonstration of diffusion involves the placement of a drop of ink in a beaker of water and watching its evolution. The measurement of surface diffusion using laser induced thermal desorption (LITD) is in this general family of concentration gradient diffusion experiments. A surface coverage gradient is established and the relaxation of the coverage gradient is monitored as a function of time.

In the LITD version of the concentration gradient method, an initial laser pulse is used to heat a well-defined area on a surface. The laser heating produces a rapid temperature jump that is large enough to desorb the adsorbates within the heated area [17-20]. Typical laser pulses have pulsewidths of tp= 100 nsec, pulse energies of 1 mJ and are focused to spatial widths of 100-200 microns (FWHM) [8,9]. Given a typical reflectivity of R=0.6 and an optical penetration depth of several hundred Angstroms for representative transition metal surfaces at λ= 1.06 micron, these laser pulses induce peak surface temperature jumps of approximately ΔT=1000-1500 K. Because only the surface region of the metal has been heated, the surface temperature will quickly cool back to the ambient surface temperature in approximately 10 tp or 1 μsec [20].

With appropriately chosen laser pulse energies, the spatial width of the desorption area will be approximately equal to the spatial width of the focused laser pulse. Under these conditions, the coverage profile at the edge of the desorption area is steep and approaches a step function [20] which facilitates a Fick's Law analysis. Moreover, the radius of these desorption areas can be measured using the spatial autocorrelation method [8]. This measurement is important because the radius of the desorption area is necessary for the Fick's Law analysis.

As a function of time, the adsorbates on the surface surrounding the desorption area will randomly walk and progressively remove the coverage gradient. Given that typical surface diffusion coefficients are generally $D < 10^{-6}$ cm^2/sec, the times required to refill the desorption area are on the order of 100 seconds or longer. Consequently, there is a clear separation between the surface heating and surface diffusion time scales.

After an appropriate delay, a second identical probe laser pulse heats the same area and desorbs the adsorbates that have diffused into the initially evacuated region from the surrounding surface. The probe laser pulse must hit the same spot with the same energy as the initial laser pulse. Otherwise, a probe desorption signal that corresponds to spatial instability rather than diffusion will be measured. Excellent spatial laser pointing stabilities can easily be obtained with the necessary precautions [9]. Unfortunately, shot-to-shot reproducibilities for laser pulse energies from Q-switched lasers are typically $\pm$ 5% at best. Consequently, this energy fluctuation can lead to spurious desorption signals. However, if several ~5 laser pulses are used to prepare and probe the desorption area, this noise on the signal caused by laser pulse energy fluctuations is minimal.

For both the prepare and probe laser pulses, the desorbed adsorbates cause a pressure transient that can be detected by a mass spectrometer. If the desorbed adsorbates are detected on their initial pass through the ionizer, a high concentration pressure burst is monitored with a fast time constant but low sensitivity in "time-of-flight" mode. [9]. Alternatively, the desorbed particles can be monitored with a slower time constant but higher sensitivity. In this mode, the desorbed particles fill the UHV chamber and a low concentration "equilibrated pressure burst" is monitored [8]. The size of the UHV chamber and the placement of the mass spectrometer will dictate which detection mode has the best signal-to-noise ratio.

The desorption flux generated by the probe laser pulse at each delay time corresponds to the amount of diffusional refilling. This desorption flux from the probe laser pulse divided by the desorption flux from the initial laser pulse defines the refilling signal [9]. The time-dependent normalized refilling signals are then fit using an equation derived from Fick's second law in order to determine the surface diffusion coefficient. The equation derived from

Fick's Law will depend on the geometry of the desorption area. Generally, either circular or elliptical desorption areas are encountered. For a circular desorption area, the time-dependent normalized refilling signal expected from Fick's Law is [16]:

$$S(t) = 1 - 2 \int_0^\infty (J_1^2(xr)/x)\exp(-Dtx^2)\,dx$$

where r is the radius of the circular desorption area and J_1 is a Bessel function of the first kind of order one.

The advantages of surface diffusion measurements using LITD techniques are their relative simplicity. The LITD method is straightforward. The identity and amount of the diffusing species is known and the data analysis is unambiguous. Likewise, the LITD method can be employed on macroscopic, single-crystal surfaces where Auger and LEED can be used to verify the surface composition and order. Likewise, the valuable methods of mass spectrometric temperature programmed desorption (TPD) can be applied to determine adlayer coverages and composition [21].

Several precautions should be taken when employing the LITD method. First, in order to reduce the surface temperature jumps required to produce adequate desorption fluxes, laser pulses of 100 - 1000 nsec should be used. These longer laser pulses will minimize the risk of surface thermal stress damage or surface melting and will not produce significant lateral surface heat migration [20]. However, even for these longer laser pulses, metals with low melting points should be avoided. Second, since the LITD methods measure surface diffusion over macroscopic dimensions of approximately 100 microns, the refilling times of $t > 100$ seconds are long enough that the desorption area can adsorb molecules from the gas phase even at UHV pressures. A check for background adsorption should be performed prior to every experiment.

Third, in order to observe diffusional refilling in a reasonable time prior to background adsorption, diffusion coefficients of $D > 10^{-6} - 10^{-9}$ cm ^{2}/sec are necessary. Given a thermally-activated diffusion mechanism, the diffusion coefficient can be tuned into this range by adjusting the temperature. However, the thermal desorption of the adsorbate may occur before diffusion coefficients of $D > 10^{-6} - 10^{-9}$ cm ^{2}/sec are reached. Consequently, for some adsorbates, the LITD method may be difficult to employ. For example, hydrogen, carbon monoxide, and various cycloalkanes and alkanes have all been successfully studied to date on Ru(001). On the other hand, water and ammonia desorb before they develop sufficient surface mobility on Ru(001).

III. Surface Diffusion of Hydrogen on Clean Ru(001)

A. Low Coverage

Using the LITD techniques described in Section II, the surface diffusion of hydrogen at low coverage was measured on clean Ru(001) [9]. Typical normalized refilling curves for three different temperatures are shown in Fig. 1. The data shown are not averages from multiple runs, but are typical results from single prepare-and-probe sequences at various delay times. Each data point corresponds to the refilling signal from separate, individual desorption areas on the macroscopic Ru(001) surface. The solid lines represent refilling curves predicted by Fick's Law corresponding to constant diffusion coefficients.

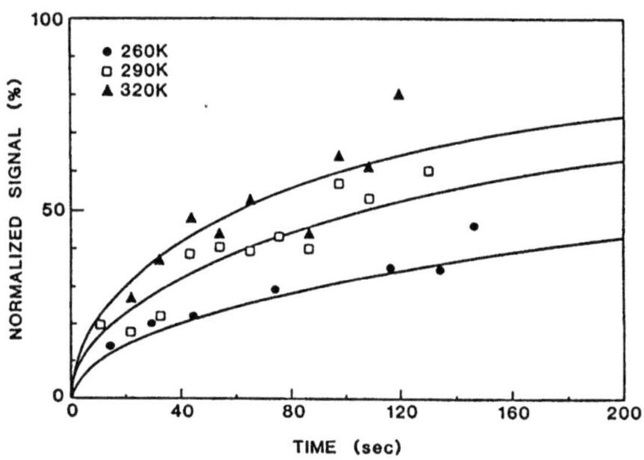

FIG. 1

The refilling curves are temperature-dependent, indicating a thermally-activated diffusion mechanism. An Arrhenius plot of the surface diffusion coefficients obtained at various temperatures is shown in Fig. 2. The measured activation barrier obtained from this plot is $E_{diff} = 4.0$ kcal/mole. Likewise, the measured preexponential factor is $D_o = 6.3 \times 10^{-4}$ cm ^{2}/sec.

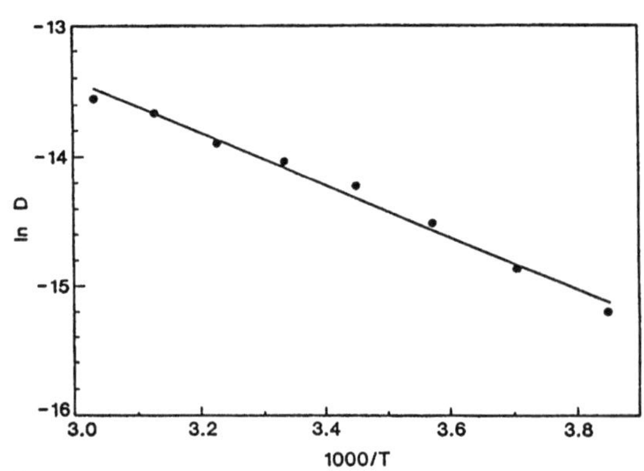

FIG. 2

Although these LITD surface diffusion experiments are performed over macroscopic dimensions, a microscopic interpretation can be obtained because these experiments are performed on single-crystal surfaces. Vibrational studies reveal that hydrogen is bound in three-fold hollow sites on Ru(001) [22,23]. Likewise, theoretical modeling demonstrates that the two-fold binding site is lower in energy than the on-top binding site. Thus the microscopic interpretation of the results is that the activation energy of 4 kcal/mole represents the surface potential barrier for hydrogen to move from one three-fold site to another three-fold site via a two-fold site as shown in Fig. 3.

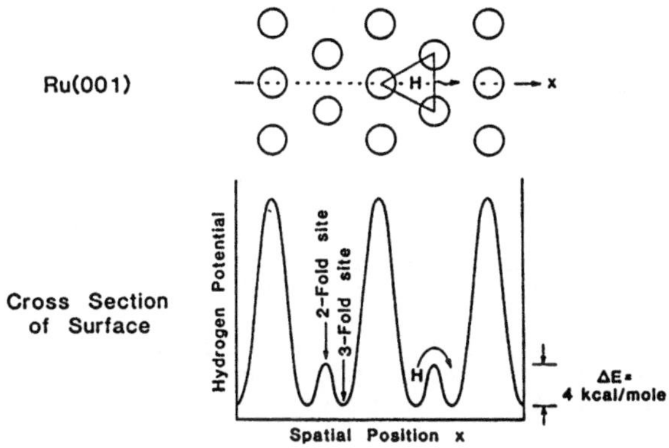

FIG. 3

In order to check the measured activation barrier and preexponential factor, the surface diffusion coefficients for hydrogen on Ru(001) were calculated using transition state theory [24]. The calculations were based on an empirical potential surface constructed using experimental values for the adsorption energy and vibrational frequencies of hydrogen on Ru(001). The calculated activation barrier was E_{diff} = 3.84 kcal/mole and the preexponential was D_o = 2.2 x 10^{-3} cm^2/sec in good agreement with the experimental results [24].

B. Effect of Coverage

The effect of coverage on the surface diffusion coefficient can reveal information regarding the adsorbate-adsorbate interactions. For example, the pioneering work of Butz and Wagner [7] on the surface diffusion of oxygen on W(110) disclosed a significant coverage dependence in the surface diffusion coefficient. This coverage dependence was attributed to adsorbate-adsorbate interactions in the oxygen overlayer [7,25]. In order to determine if hydrogen interactions on Ru(001) affect the surface diffusion coefficient, surface diffusion experiments were performed as a function of coverage [26].

A summary of the results for the hydrogen surface diffusion coefficient on Ru(001) as a function of hydrogen

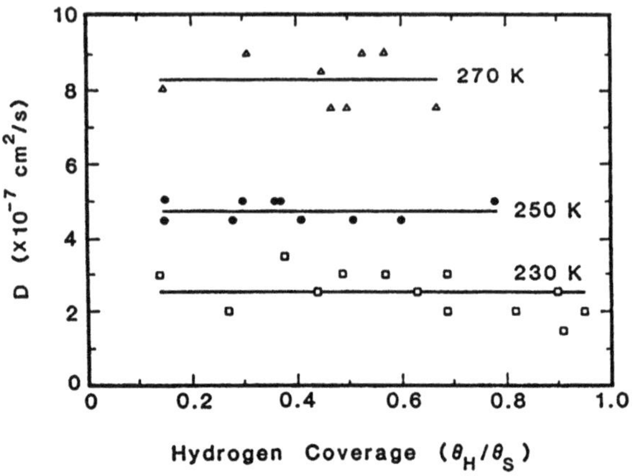

FIG. 4

coverage at three different temperatures is shown in Fig. 4. These results reveal that the surface diffusion coefficient is not coverage-dependent. The differences between the 230, 250 and 270 K results reflect the 4 kcal/mole activation barrier for hydrogen surface diffusion on Ru(001) [9].

The surface diffusion results suggest that if hydrogen adsorbate-adsorbate interactions exist, they must be much smaller than the thermal energy. This observation is consistent with previous LEED studies. No ordered hydrogen adlayer has been observed in LEED studies over a wide temperature range [27,28].

At first glance, the surface diffusion coefficient may not be expected to be independent of coverage even if adsorbate-adsorbate interactions are negligible. For example, surface diffusion may be impeded by the large number of occupied sites at high surface coverage. Since only one particle can occupy each site, the occupied sites may act to block diffusing particles. However, master equation approaches [29,30] and Monte Carlo simulations [26] have revealed that the surface diffusion coefficient is independent of coverage when site-blocking is the only interaction.

C. Isotope Effect

Quantum mechanical tunneling may influence the surface diffusion of hydrogen [31]. The isotope effect between hydrogen and deuterium surface diffusion provides a measure of the magnitude of tunneling. If the measured differences between hydrogen and deuterium surface diffusion are consistent with predictions from a thermally-activated mechanism using transition state theory, then the role of tunneling is minor.

The Arrhenius plot of the surface diffusion coefficients for hydrogen and deuterium on Ru(001) at low coverage are shown in Fig. 5 [32]. The surface diffusion

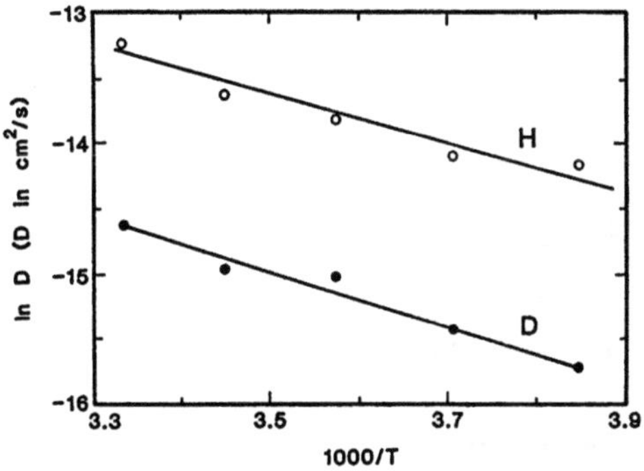

FIG. 5

coefficients for hydrogen and deuterium vary by approximately a factor of 3.5-4.0 in the temperature range from 260 to 300 K. The activation barriers were $E_{diff} = 3.6 \pm 0.5$ kcal/mole and 4.1 ± 0.5 kcal/mole for hydrogen and deuterium, respectively. The preexponential factors, given by $D = D_o \exp(E_{diff}/RT)$, were $D_o = 6.9 \times 10^{-4}$ cm ^{2}/s and 4.6×10^{-4} cm ^{2}/s for hydrogen and deuterium, respectively [32].

The measured activation barriers and preexponentials for hydrogen and deuterium are close to predictions from transition state theory [24,32]. The ratio of the preexponentials $D_o^H / D_o^D = 1.5$ is very close to the classical $m^{-1/2}$ prediction of 1.4. Likewise, the difference between the hydrogen and deuterium activation barriers of -0.5 kcal/mole is close to the predicted isotope shift of -0.23 kcal/mole obtained from transition state theory using the empirical potential energy surface for hydrogen on Ru(001) [24].

Given the similarity between the measured values and the values predicted from transition state theory, quantum mechanical tunneling constitutes a minor pathway for the surface diffusion of hydrogen on Ru(001). In the temperature range from 260 to 300 K, the surface diffusion of hydrogen and deuterium is dominated by a thermally-activated mechanism.

IV. Surface Diffusion of Hydrogen on Chemically-Modified Ru(001) Surfaces

A. Effect of Carbon

Carbon is a common surface impurity on transition metal surfaces. Likewise, surface carbon is important in various catalytic pathways on metal surfaces such as Fischer-Tropsch synthesis [33]. Because the interaction between hydrogen and surface carbon may play a key role in Fischer-Tropsch synthesis [34,35], surface diffusion studies of hydrogen on carbon-covered surfaces may provide valuable insight into this important synthetic process.

Diffusional refilling curves for hydrogen on carbon-covered Ru(001) surfaces at 300 K are shown in Fig. 6 [36].

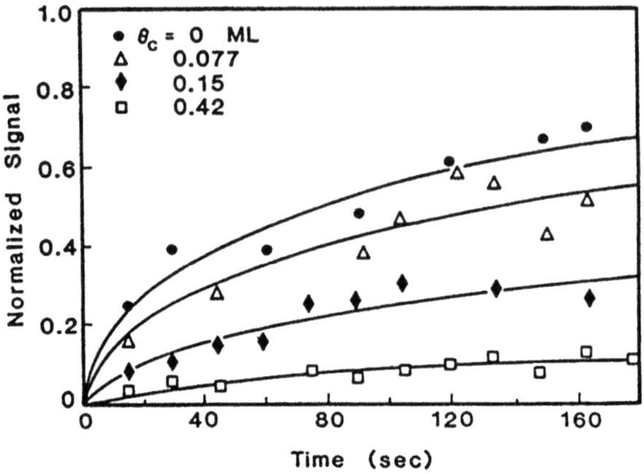

FIG. 6

The carbon coverages are given in monolayers (ML) where 1.0 ML corresponds to one carbon atom to every ruthenium surface atom. The data clearly reveals that surface carbon significantly reduces the hydrogen surface mobility.

A summary of the surface diffusion coefficients for hydrogen on Ru(001) at 300 K versus surface carbon coverage is shown in Fig. 7. The hydrogen surface mobility decreases sharply with increasing carbon coverage. For carbon coverages from θ_C = 0 to 0.42 ML, the diffusion coefficient was reduced by more than a factor of 60.

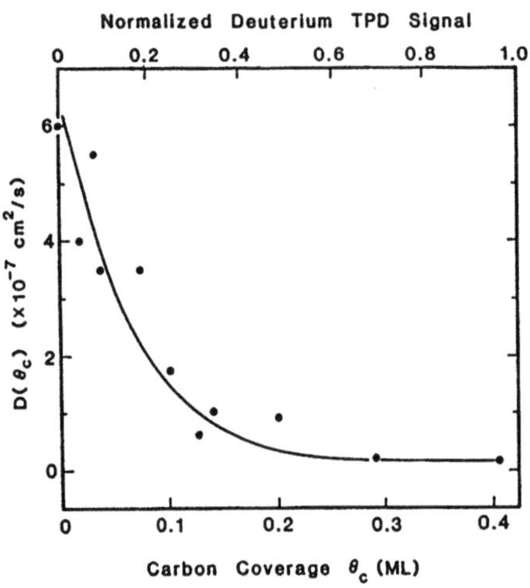

FIG. 7

Previous studies have revealed that hydrogen can interact with surface carbon to form a carbon-hydrogen bond. Post-adsorption of hydrogen on Ru(001) after deposition of carbon from ethylene decomposition gave rise to a C-H vibrational peak observed in high-resolution electron energy loss spectroscopy (HREELS) [37]. Rehydrogenation of the surface carbon was observed to occur at temperatures in the vicinity of 320-370 K.

Recent investigations also showed a marked change in the work function for hydrogen in the presence of surface carbon. In contrast to the small 25 meV positive change in the work function for hydrogen on clean Ru(001), a negative change of more then 200 meV was observed for hydrogen adsorption above 250 K on a Ru(001) surface with a small, almost immeasurable carbon coverage [28]. This change was attributed to the formation of a hydrocarbon species on the surface.

Recent studies of the Fischer-Tropsch synthesis on ruthenium have also reported that surface carbon is very reactive with hydrogen at temperatures as low as 250 K [38] and have suggested that the Fischer-Tropsch reaction proceeds by hydrogenation of a surface carbide species [33]. These observations argue that hydrogen atoms can be bound easily by surface carbon on ruthenium at 300 K. These previous experimental observations all suggest that hydrogen atoms can be easily trapped by carbon on Ru(001).

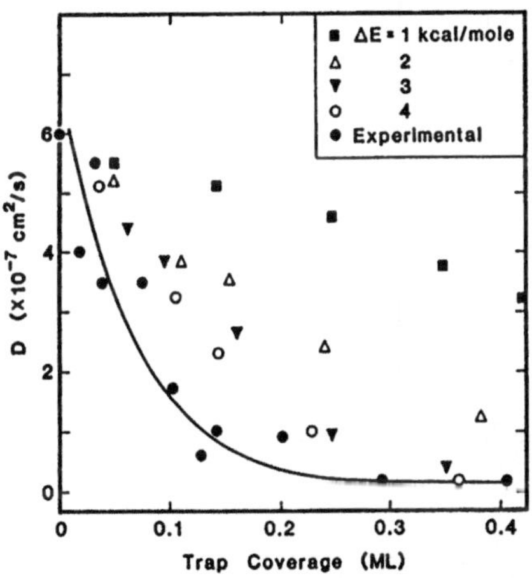

FIG. 8

Assuming a trapping mechanism, Monte Carlo simulations were performed to determine the carbon trapping energy that was consistent with the reduction of the hydrogen surface diffusion coefficient versus carbon coverage shown in Fig. 7 [36]. A comparison of the experimental data and various carbon trapping energies is shown in Fig. 8. The Monte Carlo simulations are in reasonable agreement with the data for carbon trap sites with an energy of $\Delta E \geq 3$ kcal/mole less than the normal three-fold hollow sites. The Monte Carlo simulations for the diffusion coefficient are consistently slightly higher than the data for $\Delta E \geq 3$ kcal/mole. This may indicate that some carbon traps are able to bind more than one hydrogen atom.

B. Effect of Sulfur

Sulfur is also a common impurity on catalytic surfaces. Unlike carbon, sulfur is known to block adsorption sites on surfaces either because of its large van der Waals radius [39], or because of its effect on the local surface electronic structure [40]. Surface diffusion studies of hydrogen on Ru(001) as a function of sulfur coverage were undertaken in order to explore these effects.

A summary of the surface diffusion coefficients for hydrogen on Ru(001) at 270 and 300 K versus surface sulfur coverage is shown in Fig. 9 [41]. Like the effect of surface carbon, the surface mobility of hydrogen decreased rapidly as a function of increasing sulfur coverage. At T= 300 K, the hydrogen surface diffusion coefficients dropped approximately a factor of 30 as the surface sulfur coverage increased from $\theta_S = 0$ to $\theta_S = 0.25$ ML. The differences between the results at 270 and 300 K were consistent with the 4 kcal/mole activation barrier for hydrogen surface diffusion measured earlier [9].

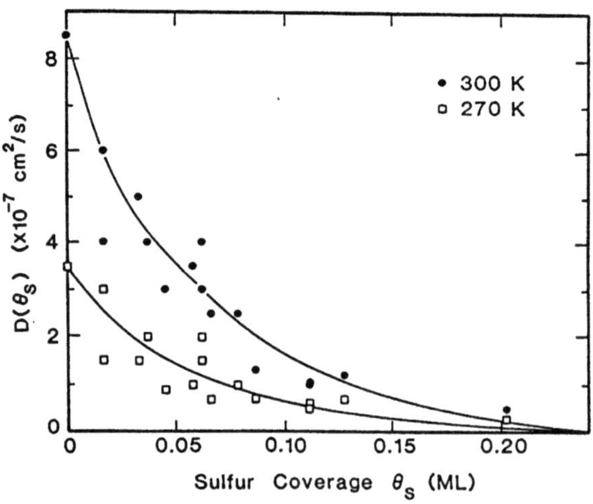

FIG. 9

Assuming a blocking mechanism, Monte Carlo simulations were performed to determine the sulfur blocking radius [41]. Simulations were performed for sulfur blocking: its own three-fold adsorption site (1 site); its own site plus its three nearest-neighbor sites (4 sites); and its own site and its three nearest-neighbor and six next-nearest-neighbor sites (10 sites). The Monte Carlo simulations are shown in Fig. 10. Agreement between the experimental data and the simulations is excellent when each surface sulfur adatom blocks ten sites.

A picture of the effect of sulfur is presented in Fig. 11. The grey region shows the van der Waals radius of a sulfur adatom on a three-fold site [39]. The three nearest neighbor three-fold sites are contained within the van der

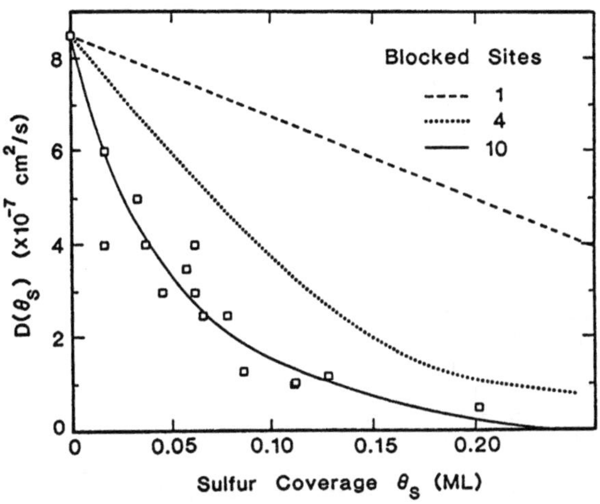

FIG. 10

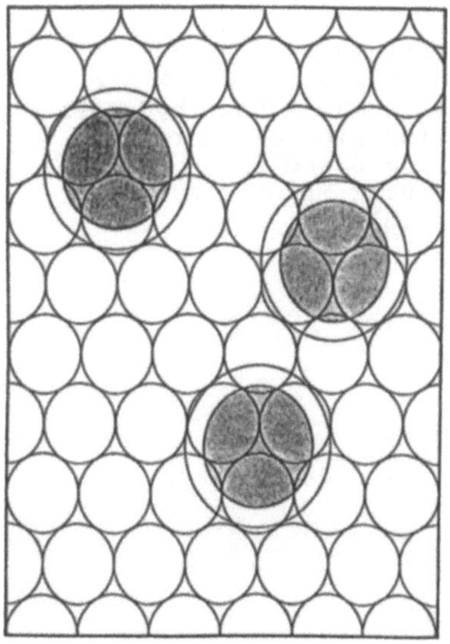

FIG. 11

Waals radius. The concentric circle shows the effective sulfur radius required to block ten sites.

The most detailed previous work on the effect of surface sulfur additives on surface kinetics has focused on hydrogen desorption, hydrogen and carbon monoxide adsorption and methanation synthesis on nickel surfaces [42-44]. This previous work has shown that the effects of sulfur can be explained if each sulfur blocks approximately 9-10 adsorption sites [43,44]. In particular, there is a striking correspondence between the decline of the methanation rate on nickel versus surface sulfur coverage [43] and the reduction of the hydrogen surface diffusion data on Ru(001) versus surface sulfur coverage shown in Fig. 9. Although these studies were performed on different transition metal surfaces, the correspondence may reveal the underlying effect of sulfur on hydrogen surface mobility which may be rate-limiting the methanation reaction.

C. Effect of Coadsorbed Carbon Monoxide

A typical chemical reaction on a surface occurs with a variety of species coadsorbed on the surface. Surface kinetics can be influenced by the interactions between surface coadsorbates. For example, coadsorbates can potentially both block surface adsorption sites and change the electronic structure of the surface. Unfortunately, very few experimental probes are capable of extricating the interactions between surface coadsorbates. Surface diffusion studies of hydrogen on Ru(001) as a function of coadsorbed carbon monoxide were conducted in order to explore these coadsorbate interactions [45].

Hydrogen surface diffusion coefficients versus coadsorbed carbon monoxide coverage at 260 and 280 K are shown in Fig. 12. The measurements revealed that hydrogen mobility was reduced slightly as a function of increasing carbon monoxide coverage between $\theta_{CO} = 0$ and $\theta_{CO} = 0.12$ ML. For carbon monoxide coverages $\theta_{CO} > 0.12$ ML, the hydrogen mobility decreased much more dramatically.

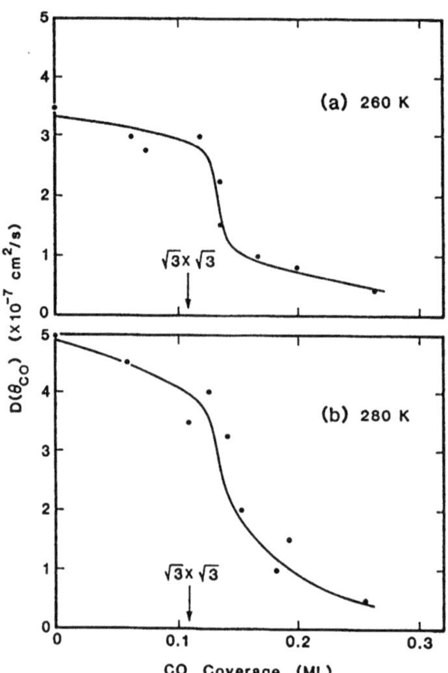

FIG. 12

Previous LEED studies have revealed that the carbon monoxide adlayer on Ru(001) undergoes an ordering transition at $\theta_{CO} \geq 0.14$ ML and T= 300 K [46]. For carbon monoxide coverages $\theta_{CO} \leq 0.14$ ML, the carbon monoxide adlayer is disordered. For carbon monoxide coverages $0.14 \leq \theta_{CO} < 0.26$ ML, the carbon monoxide adlayer consists of a partially ordered phase with short range $\sqrt{3} \times \sqrt{3}$ ordered regions. The presence of these short range $\sqrt{3} \times \sqrt{3}$ ordered carbon monoxide regions gives rise to an unmistakable and well-defined LEED pattern. LEED studies performed in this study also observe $\sqrt{3} \times \sqrt{3}$ LEED spots which first appear at a carbon monoxide coverage of $\theta_{CO} = 0.12$ ML [45]. This ordering transition occurs at $\theta_{CO} = 0.12$ ML with or without hydrogen coadsorbed on the Ru(001) surface. A picture illustrating the carbon monoxide $\sqrt{3} \times \sqrt{3}$ islands on Ru(001) at a coverage above the ordering transition is shown in Fig. 13.

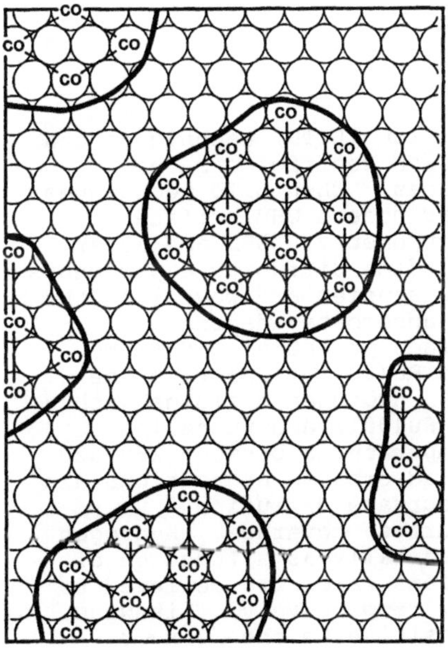

FIG. 13

Below the ordering transition at $\theta_{CO} = 0.12$ ML, the carbon monoxide adlayer is disordered and the hydrogen mobility is only weakly perturbed. This is in contrast to the trapping effect of carbon adatoms or the blocking effect of sulfur adatoms which influenced the hydrogen surface diffusion dramatically at low surface coverages. This difference may arise because carbon monoxide binds on an on-top site on Ru(001), whereas hydrogen and the surface carbon and sulfur adatoms bind in three-fold hollow sites. In addition, carbon monoxide has a much higher surface mobility than sulfur or carbon adatoms and may be able to move in response to the hydrogen adatom.

Previous studies suggest that the interaction between hydrogen and carbon monoxide on closed-packed transition metal faces is repulsive [47]. Moreover, observations of coadsorbed hydrogen and carbon monoxide on Ru(001) and Rh(111) are consistent with the segregation of hydrogen and carbon monoxide [48,49].

Consequently, after the ordering transition on Ru(001) the $\sqrt{3} \times \sqrt{3}$ ordered CO-islands form and exclude the hydrogen adatoms. This may be caused by a repulsive electronic surface potential for hydrogen in the $\sqrt{3} \times \sqrt{3}$ CO-islands or the repulsive steric interactions between CO and hydrogen which is now more noticeable because of the reduced surface mobility for CO in the ordered $\sqrt{3} \times \sqrt{3}$ structure. Therefore, in order for hydrogen to move on the surface, the hydrogen adatoms must travel around the $\sqrt{3} \times \sqrt{3}$ CO islands. This would lead to the drop in hydrogen surface mobility observed for $\theta_{CO} \gtrsim 0.12$ ML.

V. Conclusions

Laser induced thermal desorption (LITD) techniques can be used to measure surface diffusion coefficients on macroscopic, single-crystal surfaces. When LITD techniques are used in conjunction with other modern UHV surface techniques such as TPD, mass spectrometry, LEED and Auger spectroscopy, a variety of new surface diffusion experiments can be performed. These new LITD studies are revealing the underlying adsorbate-surface and adsorbate-adsorbate interactions that define the surface potential. Surface diffusion results for hydrogen on Ru(001) illustrate the potential of the LITD techniques. Hydrogen diffuses with a thermally-activated mechanism on Ru(001) and tunneling contributions are minimal for temperatures between 260-320 K. Moreover, hydrogen adsorbate-adsorbate interactions are negligible because the hydrogen surface diffusion coefficient is independent of hydrogen coverage. Hydrogen surface mobility is reduced dramatically as a function of coadsorbed carbon and sulfur adatoms which are believed to act as hydrogen traps and blocks, respectively. Likewise, hydrogen surface diffusion is sensitive to the $\sqrt{3} \times \sqrt{3}$ ordering transition in the CO adlayer when hydrogen and carbon monoxide are coadsorbed on Ru(001).

VI. Acknowledgements

This work was performed by C.H. Mak, J.L. Brand, A.A. Deckert, and B.G. Koehler, who are graduate students in the Chemistry Department at Stanford University. This work was supported by the National Science Foundation under Grant No. CHE-8514103. Some of the equipment utilized in this work was provided by the NSF-MRL Program through the Center for Materials Research at Stanford University.

References

1. G. Ehrlich and K. Stolt, Ann. Rev. Phys. Chem. **31**, 603 (1980).

2. M. Volmer and I. Estermann, Z. Physik **7** 1 (1921).

3. G. Ehrlich, Physics Today, June 1987, p. 2.

4. R. Gomer, Scientific American, August 1982, p. 98.

5. G.A. Somarjai, Chemistry in Two Dimensions: Surfaces (Cornell University Press, Ithaca, 1981).

6. D.A. King, J. Vac. Sci. Technol. 17, 241 (1980).

7. R. Butz and H. Wagner, Surf. Sci. 63, 448 (1977).

8. S.M. George, A.M. DeSantolo and R.B. Hall, Surf. Sci. **159**, L425 (1985).

9. C.H. Mak, J.L. Brand, A.A. Deckert and S.M. George, J. Chem. Phys. **85**, 1676 (1986).

10. R. Viswanathan, D.R. Burgess, Jr., P.C. Stair and E. Weitz, J. Vac. Sci. Technol. **20**, 605 (1982).

11. E.G. Seebauer and L.D. Schmidt, Chem. Phys. Lett. **123**, 129 (1986).

12. D.R. Mullins, B. Roop and J.M. White, Chem. Phys. Lett. **129**, 511 (1986).

13. B. Roop, S.A. Costello, D.R. Mullins and J.M. White, J.Chem. Phys. **86**, 3003 (1987).

14. M.A. Vannice, J. Catalysis **37**, 449 (1975).

15. M.A. Vannice, J. Catalysis **37**, 462 (1975).

16. J. Crank, The Mathematics of Diffusion, 2nd Ed. (Oxford Univ. Press, London, 1975)

17. G. Wedler and H. Ruhmann, Surf. Sci. **121**, 464 (1982).

18. R.B. Hall and A.M. DeSantolo, Surf. Sci. **137**, 421 (1984).

19. J.P. Cowin, D.J. Auerbach, C. Becker and L. Wharton, Surf. Sci. **78**, 545 (1978).

20. J.L. Brand and S.M. George, Surf. Sci. **167**, 341 (1986).

21. D.A. King, Surf. Sci. **47**, 384 (1975).

22. A. Barteau, J.Q. Broughton and D. Menzel, Surf. Sci. **133**, 443 (1983).

23. H. Conrad, R. Scala, W. Stenzel and R. Unwin, J. Chem.Phys. **81**, 6371, (1984).

24. C.H. Mak and S.M. George, Chem Phys. Lett. **135**, 381 (1987).

25. M. Bowker and D.A. King, Surf. Sci. 72, 208 (1978).

26. C.H. Mak, J.L. Brand, B.G. Koehler and S.M. George, Surf. Sci. 191, 108 (1987).

27. H. Shimizu, K. Christmann and G. Ertl, J. Catalysis **61**, 412 (1980).

28. P. Feulner and D. Menzel, Surf. Sci. **154**, 465 (1985).

29. R. Kutner, Phys. Lett. **81A**, 239 (1981).

30. H.J. Kreuzer, This Volume.

31. R. DiFoggio and R. Gomer, Phys. Rev. B **25**, 3490 (1982).

32. C.H. Mak, J.L. Brand, B.G. Koehler and S.M. George, Surf. Sci. 188, 312 (1987).

33. P. Biloen and W.M.H. Sachtler, Advan. Catalysis **30**, 165 (1981).

34. D.L. King, J. Catalysis **61**, 77 (1980).

35. J.C. Ekerdt and A.T. Bell, J. Catalysis **58**, 170 (1979).

36. C.H. Mak, J.L. Brand, B.G. Koehler and S.M. George, J. Chem. Phys. 87, 2340 (1987).

37. M.A. Barteau, J.Q. Broughton and D. Menzel, Appl. Surf.Sci. **19**, 92 (1984).

38. H. Wise and J.G. McCarty, Surf. Sci. **133**, 311 (1983).

39. L. Pauling, Nature of the Chemical Bond, 3rd Ed. (Cornell University Press, Ithaca, 1960).

40. P.J. Feibelman and D.R. Hamann, Phy. Rev. Lett. **52**, 61 (1984).

41. J.L. Brand, A.A. Deckert and S.M. George, "Effect of Sulfur on the Surface Diffusion of Hydrogen on Ru(001)," Surf. Sci. (in press).

42. S. Johnson and R.J. Madix, Surf. Sci., **108**, 77 (1981).

43. M. Kiskinova and D.W. Goodman, Surf. Sci. **108**, 64 (1981).

44. W. Erley and H. Wagner, J. Catal. **53**, 287 (1978).

45. C.H. Mak, A.A. Deckert and S.M. George, "Effect of Coadsorbed CO on the Surface Diffusion of Hydrogen on Ru(001)," (submitted to J. Chem. Phys.)

46. H. Pfnur and D. Menzel, Surf. Sci. **148**, 411 (1984).

47. J.M. White, J. Phys. Chem. **87**, 915 (1983).

48. D.E. Peebles, J.A. Schreifels and J.M. White, Surf. Sci. **116**, 117 (1982).

49. E.D. Williams, P.A. Thiel, W.H. Weinberg and J.T. Yates, Jr., J. Chem. Phys. **72**, 3496 (1980).

Reflection High Energy Electron Diffraction Studies of Diffusion and Cluster Formation During Molecular Beam Epitaxy

P.R. Pukite, G.S. Petrich, G.J. Whaley, and P.I. Cohen

Department of Electrical Engineering,
University of Electrical Engineering, University of Minnesota
Minneapolis, MN 55455, USA

Understanding the surface processes controlling molecular beam epitaxy (MBE) is crucial to the growth of high quality epitaxial structures. The fundamental mechanisms involved in the growth include adatom and impurity adsorption, surface reconstruction, surface diffusion, cluster formation, step propagation, and evaporation. Reflection high-energy electron diffraction (RHEED) is useful because it probes the surface reconstruction and the step structure during the growth process. In particular, the competing processes of cluster formation and step propagation are followed as a function of growth conditions. The growth of GaAs and Ge by MBE has been investigated in detail with RHEED. Layer-by-layer growth is observed on these surfaces over a range of conditions. In both systems, the growth mode changes from cluster formation at low substrate temperatures to step propagation at high temperatures. The underlying step morphology also affects the growth mode. By introducing an array of steps due to a misorientation from a low-index plane, step propagation is strongly enhanced. The step motion and stability is interpreted within the framework of several growth models. The models give insight into why the MBE layers can be grown so smoothly.

1. Introduction

The growth of GaAs and other III-V materials by molecular beam epitaxy (MBE) has provided precise control for the fabrication of layered structures in many opto-electronic applications. This control is particularly important at the heterojunction interfaces incorporated in many of the devices. For example, the quality of an interface is influenced by the transition between dissimilar materials and the smoothness of the initial surface. MBE is able to meet these requirements as it is a low temperature process that employs fast switching (relative to the growth rate) between constituent fluxes, thereby achieving abrupt interfaces with limited interdiffusion between the layers. If the growth of the layers can be better understood, the quality of ultrathin superlattices and metastable structures will likely improve.

Reflection high-energy electron diffraction (RHEED) is an *in situ* surface sensitive tool that allows instant feedback on the growth process. The most fundamental information that the RHEED pattern contains regards the lattice constant and any surface reconstruction that may exist. Additionally, information on the morphology of the surface is obtained from the shape of the diffracted beams. At one limit, a very rough surface containing three-dimensional asperities gives a transmission-like diffraction pattern. At the other extreme, smooth, stepped surfaces exhibit RHEED patterns with short streaks that have structure along their length. By analyzing the shape of these streaks, the nature of the steps on the surface can be understood.

In the following section, we will discuss the mechanisms of crystal growth as a basis for interpreting the RHEED measurements. The roles of adatom adsorption, surface diffusion, cluster formation, and step stability will be introduced here. In the third section, diffraction from stepped surfaces will be explained. In the final section, RHEED measurements from growing surfaces of GaAs(100) and Ge(100) will be presented. The ultimate goal is to understand the interactions between adatoms and steps that lead to the growth of smooth layers.

2. Mechanisms of Crystal Growth

2.1 Adsorption

The first stage of growth from the vapor phase is adsorption of the atomic or molecular species onto the crystal surface. In this state the adatom can be considered weakly bound due to its low coordination. For an elemental material like Ge, both the gaseous and adsorbed species exist as a monomer. In the case of GaAs, little is known about the adsorbate stage. Pioneering work done by Arthur [1] and Foxon and Joyce [2] has provided an initial understanding of the adatom species involved. From these studies, it is well understood that Ga exists as a monomer and is the rate controlling step during the growth. Arsenic is believed to exist in a precurser dimer state $As_2{}^*$ that is weakly bound and which will reevaporate unless sufficient Ga is present. The classical argument is that crystal growth proceeds as the weakly bound adatoms migrate to regions of higher coordination such as kink sites at step edges [3].

A complicating issue is the presence of surface reconstruction. The reconstruction is thought to stabilize the growth front [4,5]. For example, the bulk-terminated (100) face of diamond will not change surface free energy (number of dangling bonds) with the introduction of a growth unit, neglecting next nearest neighbor interactions (i.e. reconstruction) [4]. Therefore, such a surface should grow atomically rough. However, the reconstruction of a (100) surface will lead to an increased surface energy if adatoms disrupt the local bonding arrangement and form isolated growth units. Under these circumstances, the free energy change is minimized by incorporating atoms at only the step sites. For reconstructed surfaces such as (100) GaAs and Ge, the lowest energy growth process is believed to be step propagation from clusters or substrate steps. For other faces such as (111) diamond and the (100) simple cubic used in growth simulations below the roughening temperature [36], reconstruction is not needed to force step propagation.

For Ge(100) a 2×1 reconstruction is observed over a wide temperature range. Because of two possible domains, this reconstruction is generally observed as a 2×2 diffraction pattern. The 2×1 is considered to result from pairing or orbital rehybridization of adjacent atoms in the dangling bond direction.

An introduction of a growth unit requires this bonding to be broken [6].

GaAs(100) exhibits several reconstructions. These can be separated into As-stabilized and Ga-stabilized depending on whether the surface is As rich or As deficient. The reconstructions can be obtained by varying As flux or substrate temperature independently [7]. Typically, growth is performed under As-stabilized conditions and a 2×4 reconstruction. In this case, the 2× is thought to be (at least to first order) a pairing of adjacent As atoms similar to that of Ge.

2.2 Surface Diffusion

The rate at which the adatoms reach low-energy step edges and kinks is dependent on surface diffusion. Burton, Cabrera, and Frank (BCF) [3] included diffusion in a model that describes crystal growth solely by the propagation of surface steps. Within this model, the concentration gradient of adatoms toward step edges is considered the driving force for step propagation. The solution to the BCF diffusion equation includes the approximation of low-supersaturation or close to equilibrium growth. This is far removed from the conditions typically used during MBE [8]. Several groups [9,10,11] have modified the equations to take into account the high supersaturation conditions employed during MBE. The major alteration involves including a moving reference frame into the analysis.

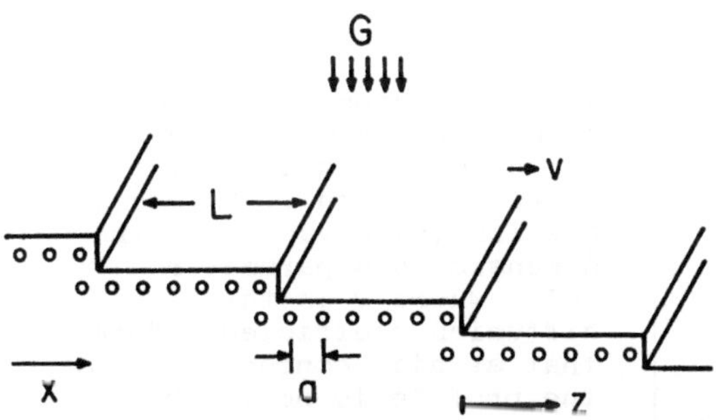

Fig. 1. Step train or staircase of steps. The step velocity is denoted by v and the position along a terrace relative to the step edge by z. In the modified BCF model, growth proceeds as adatoms impinge on the surface in a uniform flux, G, and attach to step edges.

Figure 1 shows the stepped surface or staircase used in the modified BCF analysis. Here, surface steps of fixed height separate low-index terraces of uniform length. The step height is generally a multiple of the interplanar spacing while the terrace length, L, is determined by the specific surface misorientation. We neglect surface evaporation, which is a good approximation for both GaAs and Ge over a certain temperature range. Then the diffusion equation can be written as,

$$\frac{dn(x,t)}{dt} = D\,\frac{d^2n(x,t)}{dx^2} + G/a \quad , \tag{1}$$

where n(x,t) is the concentration of adatoms, D is the chemical diffusion coefficient, G is the growth rate in monolayers/time, and a is the lattice constant. At the step edges the equilibrium adatom concentration is taken to be zero. The assumption here is that all adatoms that reach the step edge incorporate. The inclusion of the moving frame, $z = x - \int v(t)dt$, where v(t) is the step velocity modifies the equation to the following form,

$$\frac{dn(z,t)}{dt} = D\,\frac{d^2n(z,t)}{dz^2} + v(t)\,\frac{dn(z,t)}{dz} + G/a\,. \qquad (2)$$

This is a nonlinear time-dependent differential equation, which requires numerical methods to solve [16]. In the steady state, however, where $v(\infty) = GL$, the solution is straightforward.

$$n(z,\infty) = n_0\cdot\left\{\frac{1-\exp(-zLG/D)}{1-\exp(-L^2G/D)} - \frac{z}{L}\right\} \qquad (3)$$

Here n_0 is the one dimensional surface site density, 1/a. In this model the velocity of the step edges is proportional to the gradients feeding them as follows,

$$v(t) = aD\cdot\left\{\frac{dn(z=0,t)}{dz} - \frac{dn(z=L,t)}{dz}\right\}\,. \qquad (4)$$

Figure 2 shows adatom concentration profiles for various values of L^2G/D.

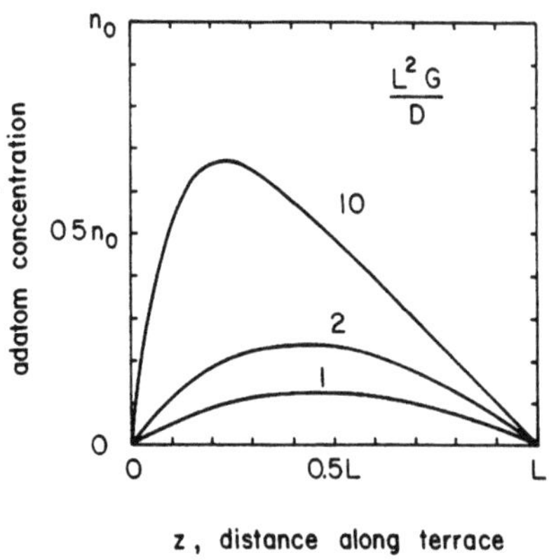

Fig. 2. Adatom concentration along terrace of length L during steady state growth. Curves labelled 1, 2, and 10 are for these values of the dimensionless parameter L^2G/D where D is the diffusion coefficient. Note that at high concentrations, the profile is noticeably asymmetric. This stems from the step motion.

Several items should be noted at this point. First, the effect of cluster formation has not been included in the calculation. Second, the profiles are asymmetric. These are addressed next.

To understand the significance of cluster formation, consider that for $L^2G/D < 1$ the maximum adatom concentration is given as,

$$n_{max} \approx \frac{L^2G}{8Da}\,. \qquad (5)$$

As $L^2 G/D > 1$, the peak adatom concentration approaches 1/a (which is the limit of available sites), and the adatoms are not free to hop from site to site without making frequent collisions with other adatoms. At this point, the above equations must be modified to account for the high probability of near neighbors [12]. A coverage dependent diffusion coefficient may be justified [37].

To understand the asymmetry, note that since $G = 1/\tau$, where τ is the monolayer time, a natural diffusion length can be expressed as $r = \sqrt{2D\tau}$. Equation 5 then can be written as the square of the ratio between L/2 and diffusion length. The diffusion length gives the mean distance an adatom can travel in time τ. If the diffusion length is less than L/2, there is a high probability for the adatoms to remain in the terrace region and be swept up by the propagating step instead of diffusing toward it. This is called growth by convection [11]. For progressively larger diffusion lengths, the adatoms will more likely diffuse toward a step. Then $L^2/8D$ gives a mean incorporation time for an adatom which lands near the middle of a terrace.

To summarize, the calculation of adatom concentration within this model shows the interrelation of growth rate, diffusion length, and step density that must clearly extend to the MBE environment. The clear challenge is to determine the microscopic contributions to D under realistic MBE conditions. The preceding macroscopic treatment of equilibrium chemical diffusion must be modified to include the non-equilibrium environment that both isolated adatoms and clusters experience during growth. Until that analysis is available, this equilibrium description should serve as a starting point to understand the data.

2.3 Cluster Formation

In the previous section, the possibility of cluster formation was mentioned. In general one would expect that an increase in adatom concentration would lead to an increased probability of cluster formation. The probability increase can be thought of as an increased collision rate between adatoms in a high concentration region,

$$R_{coll} \approx n^2 \overline{V} \approx n^2 D/a \quad , \tag{6}$$

where $\overline{V}$ is the mean velocity of adatoms and R_{coll} is the collision rate, neglecting the scattering cross section [9,13]. However, the collision of two adatoms may or may not lead to a growing cluster. The stability of a cluster is dependent on temperature. At high temperatures, the dissociation rate of clusters may be greater than the collision rate between adatoms [9,14]. This points out the difficulty of separating temperature dependent kinetic processes such as diffusion and cluster dissociation. Additionally, the rates of cluster formation and diffusion may depend on the addition of impurities to the surface. In this case, impurities can occupy preferred step sites, causing increased reflectivity of adatoms at steps or an increase in collision rate on the terrace.

Perhaps the strongest effect on cluster formation is the separation of steps. Combining Eqs.(5,6), one can see that the collision rate in the middle of a terrace increases as L^4. We will investigate this more thoroughly in the results section.

2.4 Stability of Propagating Steps

We have assumed that the staircases are highly ordered (all terraces of equal length) in calculating the adatom concentrations within the modified BCF theory. In fact the motion of adatoms toward the steps stabilizes this step train against fluctuations. In Fig.1, the staircase is considered stable against fluctuations if the number of adatoms collected at a step is greater from the lower terrace than from the upper terrace. This can be shown mathematically from a normal mode analysis [15].

The physical reason for this stability can be understood if a fluctuation giving rise to an isolated long terrace were to occur. In this situation, since a long terrace increases the asymmetry of the concentration profile in Fig.2, the preceding step (to the left of the terrace in Fig.1) will advance more rapidly due to convective growth than the step in front of the terrace. This will tend to decrease the length of the long terrace. Alternatively, from Eqs.(3,4) the current densities to the left and right of a terrace can be calculated, which results in more adatoms being collected from the step preceding the terrace. The result of this is that during growth, the staircase should remain well ordered.

3. Diffraction from Stepped Surfaces

3.1 Misoriented Surfaces

The RHEED pattern from a stepped surface contains some distinguishing characteristics. Chief among them is a broadening or splitting of the diffracted beams. In particular, diffraction from an ordered array of steps such as that shown in Fig.1 causes a sharp splitting in final angle of the specular beam [17]. This can be understood by considering that waves diffracted from adjacent terraces can either constructively or destructively interfere. Under the condition of destructive interference, the splitting is the sharpest and clearest. The corresponding separation between the split beams is inversely related to the average terrace length. If the beam is directed down the array of steps, as in Fig.3, the angular separation is increased by $1/\sin\theta_i$, where θ_i is the incident angle.

If the terrace lengths are disordered, the splitting will not be as sharp and the individual components will broaden [18]. The broadening can be related to a rms fluctuation in the terrace lengths if the lengths are statistically independent. Similarly if random up and down steps, due to clusters, are formed on the terraces, the peak shape will broaden [18,19].

3.2 Surfaces with Two-Dimensional Clusters

The diffraction pattern from a surface devoid of steps consists of sharp beams at each of the integral-order beam positions. The

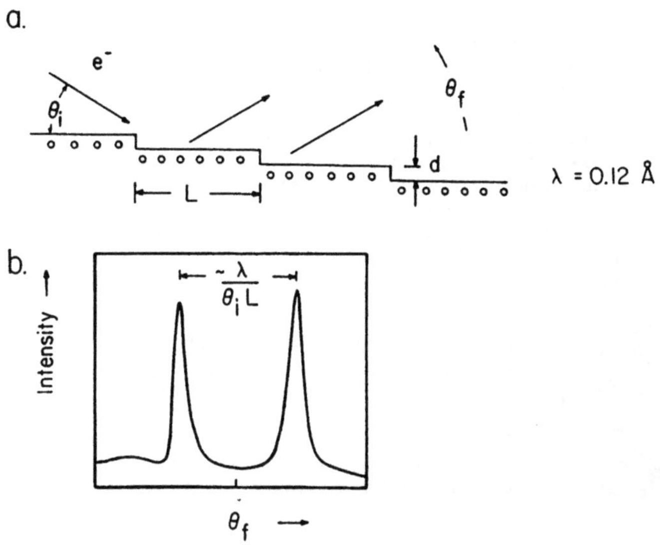

Fig. 3. Incident RHEED geometry for a surface staircase is indicated in (a). If incident angle is set such that waves scattered from adjacent terraces destructively interfere, the diffracted beam(s) will split as shown in (b).

sharpness of the beams is limited by the incident beam optics and by any long-range correlations on the surface. With the introduction of up and down steps due to clusters, short-range correlations are increased and each of the integral order beams will broaden. This is shown in Fig.4. Lent and Cohen [20] calculated the exact diffracted beam profiles for a one-dimensional geometric distribution of steps. By careful fitting of the calculated profiles to measurements, one can extract both surface coverage and mean correlation length (~ average island separation) on the surface [20]. In this paper, we would like to demonstrate another powerful capability of RHEED, namely that of determining cluster shapes from beam profiles.

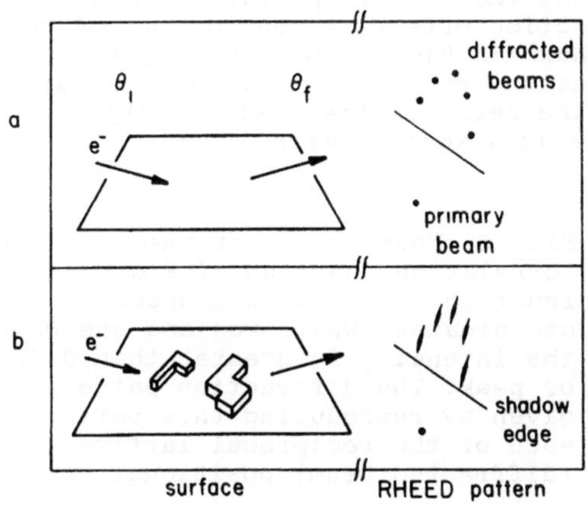

Fig. 4. (a). RHEED pattern from a step-free surface. Each of the diffracted beams is sharp. The primary beam is due to the incident beam that misses the sample and the shadow edge defines the low-angle takeoff of scattered electrons. (b). As two-dimensional clusters are formed on the surface, diffracted beams streak in a direction normal to the surface. This is due to an increase in the short-range correlations on the surface.

To calculate the diffraction from an arbitrary distribution of two-dimensional clusters requires a knowledge of the two-dimensional correlation function. The diffracted intensity is then the Fourier transform of this function. However, the two-dimensional correlation function is in general difficult to calculate [20]. To make the calculation more tractable, a simple system is analyzed as shown in Fig.5. In this example, the clusters are square in shape and arranged in a checkerboard configuration. What makes the intensity distribution particularly easy to calculate here is that we assume random or incoherent sums of phases for the various clusters.

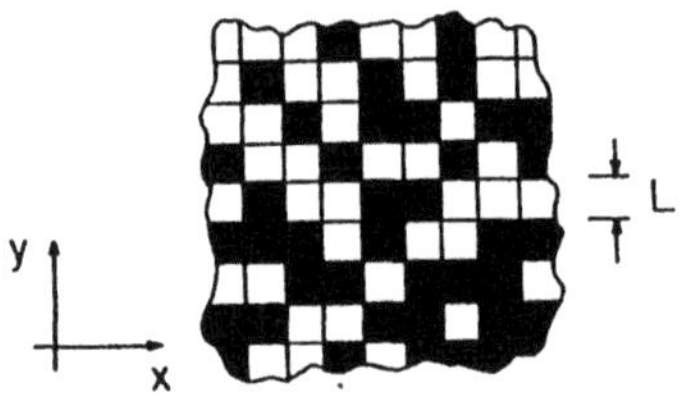

Fig. 5. Checkerboard configuration of square clusters. The squares indicate clusters that have statistically independent values of reflectance. In this example, they are randomly assigned either black or white by flipping a coin.

The statistical independence can be realized if there are many levels on the surface and the step heights are uncorrelated from cell to cell. The correlation function for the surface is then just the correlation function of a single cell,

$$C(x,y) = \left[1 - \frac{|x|}{L}\right]\cdot\left[1 - \frac{|y|}{L}\right] , \tag{7}$$

where C(x,y) is the two-dimensional correlation function and L is the cell size. The Fourier transform of this gives the diffracted intensity.

$$I(S_x, S_y) = L^2 \cdot \frac{\sin^2(LS_x/2)}{(LS_x/2)^2} \cdot \frac{\sin^2(LS_y/2)}{(LS_y/2)^2} \tag{8}$$

Here, S_x and S_y are the reciprocal space coordinates. The most interesting feature of this intensity distribution is that most of the intensity is located along the two reciprocal space axes. This is a purely geometric diffraction effect. A contour map of the intensity distribution is shown in Fig.6. In a real system, disorder in the sizes of clusters would be expected to occur. This would tend to smoothen out the zeros in the shape in Fig.6. However, the tails along the axis would remain.

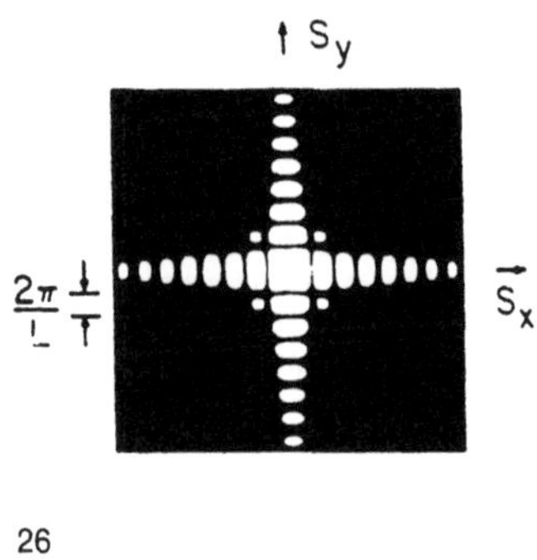

Fig. 6. Fourier transformed correlation function of Fig.5 as a function of reciprocal space coordinates. White regions are where the intensity is greater than 0.001 of peak. The diffraction pattern is given by reproducing this pattern at each of the reciprocal lattice (diffracted beam) positions.

The transformation of the intensity map to a diffraction pattern is straightforward when applied to LEED. In this case, the diffraction pattern would be exactly Fig.6. The spot shape can then be considered as a finite size effect and a direct comparison to experiment can be made. However in RHEED, the transformation is more complex. If one is to compare to experiment (as will be done in Sec. 4.2), the RHEED transformed intensity map must be calculated.

As a rule the glancing geometry of RHEED will elongate the distribution of Fig.6 along the direction of the incident beam. For example, if the electron beam was oriented along the x axis of the clusters, the projected diffraction pattern would show the profile along S_x to elongate. To determine what the expected beam shape should be requires the following equations of projected momentum transfer.

$$S_x = k\cdot(\cos\theta_f \cos\Phi_f - \cos\theta_i \cos\Phi_i) \quad (9)$$

$$S_y = k\cdot(\cos\theta_f \sin\Phi_f - \cos\theta_i \sin\Phi_i) \quad (10)$$

Here, $k = 2\pi/\lambda$ is the electron wavevector, θ_f and θ_i are the glancing final and incident angles of the electron beam, and Φ_f and Φ_i are the final and incident azimuthal angles of incidence. When $\Phi_i = 45°$, the electron beam is directed along the diagonal of a cell. Figure 7 shows the RHEED projected intensity distribution at this azimuth for an angle of incidence of 28 mrad. Here the streaking is normal to the surface much like Fig.4b, but with structure resulting from intensity lying along the two reciprocal space axes. In the results section, we will compare this plot to measurements.

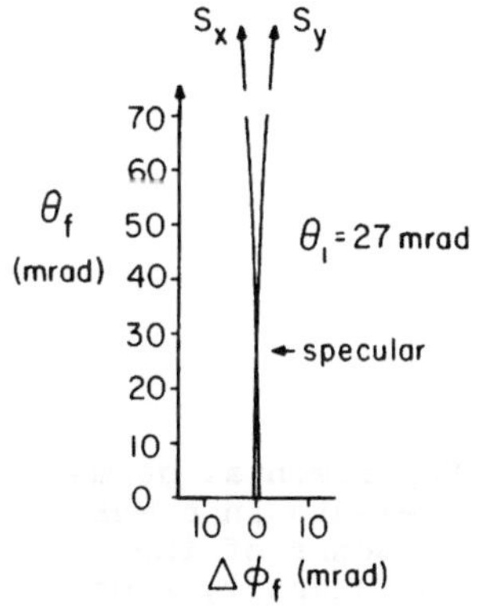

Fig. 7. Calculation of RHEED pattern of Fig.6 when beam is directed diagonally across a square cluster. Tails along S_x and S_y elongate in a direction normal to the surface as shown by the slightly curved lines. These meet at the specular beam position.

3.3 RHEED Intensity Oscillations

Crystal growth on a surface without substrate steps would be expected to proceed as cluster formation followed by step propagation of the cluster edges. As the clusters meet and coalesce, a smooth surface is recovered. The next layer would repeat the cycle. If a RHEED beam is monitored during this process the intensity would be observed to follow a similar cyclic behavior of monolayer period. This is the basis for RHEED intensity oscillations [21,22,23].

The simplest explanation for the diffracted intensity behaviour was outlined in the previous section. The diffracted peak intensity is the highest for a smooth surface. The growth is initiated when the source shutter is opened. At this point, growth clusters are formed on the surface. The diffracted beam broadens and decreases in intensity due to the decrease in correlation length on the surface. At 1/2 coverage in a two level system, the surface roughness reaches its maximum and this corresponds to the minimum value in peak intensity. Beyond this point, the surface becomes smoother and the peak intensity recovers. The intensity oscillations have been observed in many growth systems, including metals [24,25]. In the next section, we will use the oscillations in conjunction with the other diffraction analyses presented earlier to understand the growth of GaAs and Ge(100).

4. Results of Growth on GaAs and Ge(100)

4.1 Experimental

Epitaxial GaAs films were grown on GaAs(100) wafers in a Physical Electronics MBE 400 system. Details of the growth can be found elsewhere [26]. Ge films were grown in the same system using either GaAs or Ge(100) substrates. On the GaAs substrates, Ge buffer layers of 1000 Å were grown before measurements were made. Misorientations were measured by x-ray $\theta-2\theta$ goniometry. A schematic of the diffraction apparatus is shown in Fig.8. The

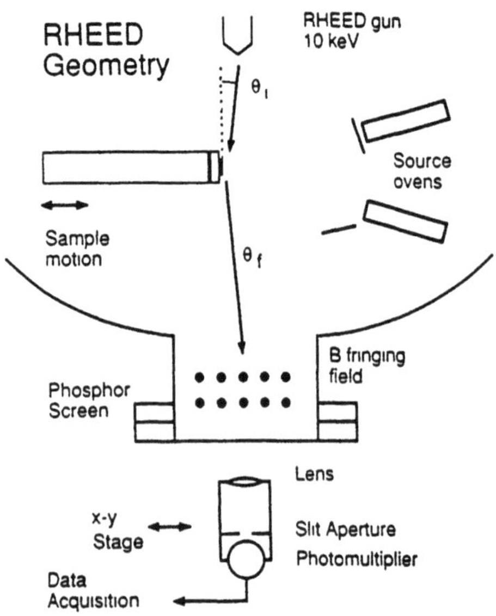

Fig. 8. Experimental geometry for measurements in a MBE apparatus. Scans of the diffracted intensity along θ or ϕ are obtained by magnetically deflecting the diffracted beams across a stationary detector. From the shape of these profiles, e.g. Fig.3, the surface morphology is characterized.

diffractometer consists of a 10 keV electron gun that is focussed at the phosphor screen. The light from the phosphor screen was detected by a slit/photomultiplier assembly. Angular profiles of the diffracted beam were obtained by moving the diffraction pattern with a pair of electromagnetic coils. The glancing incident angle could be varied by a combination of sample motion and beam deflection. The incident azimuth could be varied by rotating the sample about the surface normal. The scattering angles were determined by measuring the azimuthal position of the sample relative to symmetry directions and from the separation between the specular beam, primary beam, and shadow edge.

4.2 GaAs growth

Since the (100) is not a cleavage face, all surfaces with this orientation are expected to contain steps. This is mainly a result of the difficulty of orienting, cutting, and polishing with the aim of eliminating steps entirely. One should realize that even a misorientation of 1 mrad from the (100) will contain on the average 3000 Å between surface steps. This is within the region that RHEED is sensitive to steps. The RHEED studies of GaAs growth on such surfaces has focussed on the shape of the diffracted beam as a function of growth conditions [27,28].

The main result of these measurements has been a determination of the correlation length of clusters during the initial stages of growth. The correlation length was determined at nearly half coverage, i.e. at the minimum of an intensity oscillation. Here, the full-width at half maximum of the angular profile is inversely related to the correlation length. The correlation length was observed to increase as a function of temperature, particularly as the surface reconstruction passed from the 2×4 to the "1×1" reconstruction [28]. The activation energy associated with the "1×1" correlation length temperature dependence was 4 eV, which was much higher than the 2×4 dependence. The implication of this was that some non-diffusion bond breaking process was occurring. Other measurements have indicated that evaporation of GaAs is important only in the high temperature "1×1" reconstruction [18,29].

These measurements have been extended to misoriented surfaces. Figure 9 gives a comparison of growth oscillations on a 1 mrad and a 35 mrad misorientation. The lack of strong oscillations on the 35 mrad surface has been attributed to the high density of step sites available for step propagation [19]. We have measured the step height by RHEED to be 2.85 Å = a/2. For the 35 mrad surface, the average terrace length is 80 Å. Recalling that the average number of collisions in the middle of a terrace goes as L^4, this number can be reduced by a factor $\sim10^6$ by reducing the average terrace length from 3000 Å to 80 Å. This reduction is directly observed in the weakness of the oscillations. In the limit of pure step propagation, no oscillations should be observed as there is no cyclic variation in roughness. The weak oscillations observed on the 35 mrad surface are from the few adatoms that form critical clusters before reaching the propagating step edges. We should note that sharp splitting is maintained throughout the growth. This implies that the staircase does not undergo severe

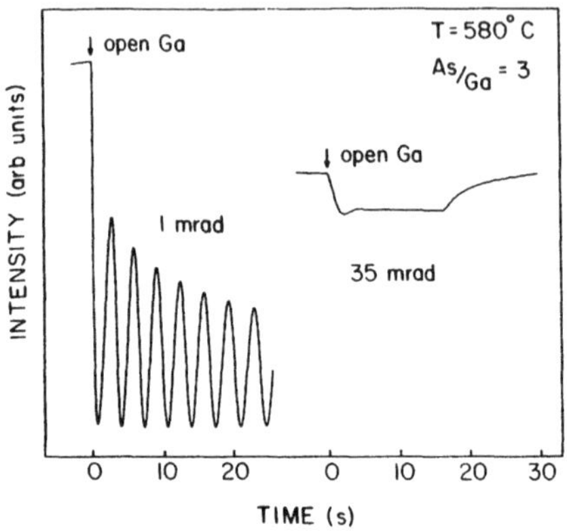

Fig. 9. Temporal oscillations in the RHEED intensity during the growth of GaAs(100) by MBE. Strong oscillations are observed when the surface is within 1 mrad of the (100) plane. Weak oscillations are observed under the same conditions if the surface is misoriented by 35 mrad toward the (111)B plane. This indicates less cluster formation when a high density of substrate steps are present. The incident angle is 30 mrad and the beam is directed perpendicular to the staircase.

fluctuations in terrace length distribution and perhaps an ordering mechanism similar to that described in Sec.2.4 is operative.

Neave and coworkers [30] have tried to estimate the diffusion coefficient by adjusting the growth rate and substrate temperature until the oscillations completely disappear on a 2.5° surface. In effect this modifies the adatom concentration according to Eq.5. Since the diffusion coefficient is a temperature activated process, the goal is to record the growth rate needed to barely detect oscillations as a function of temperature. A one-to-one correspondence can then be set up between growth rate and diffusion coefficient at each temperature. We have extended these measurements over a larger temperature range.

Figure 10 is a plot of the temperature needed to completely extinguish oscillations as a function of growth rate. At low temperatures within the 2×4 reconstruction the Neave analysis may apply, however near the transition to the "1×1" reconstruction the cutoff point is insensitive to the growth rate. This again implies that a different bond breaking process may be occurring.

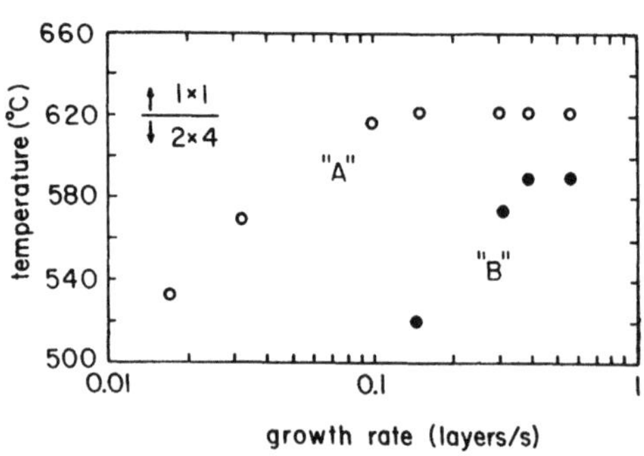

Fig. 10. Highest temperature needed to completely extinguish intensity oscillations as a function of growth rate for (100) surfaces misoriented by 35 mrad toward (111)A (open circles) and (111)B (closed circles). A broader temperature range for the (111)A indicates more clustering on this surface misorientation.

However, we should note that the direction of misorientation is important in this analysis. In Fig.10, two types of step edges are measured. If the surface is misoriented in the four-fold direction, As terminated step edges are revealed. If the surface was misoriented 90° to this, in the two-fold direction, the step edges are Ga terminated. For the misorientation with Ga terminated steps, much stronger oscillations were observed. In fact, the As terminated stepped surface often exhibited no oscillations over the entire temperature range. A plausible explanation is that the diffusion rate is higher in the [0$\bar{1}$1] direction, i.e. the direction of the two-fold reconstruction. If this were the case, the diffusing adatoms have a lower probability of cluster formation before reaching the As terminated steps.

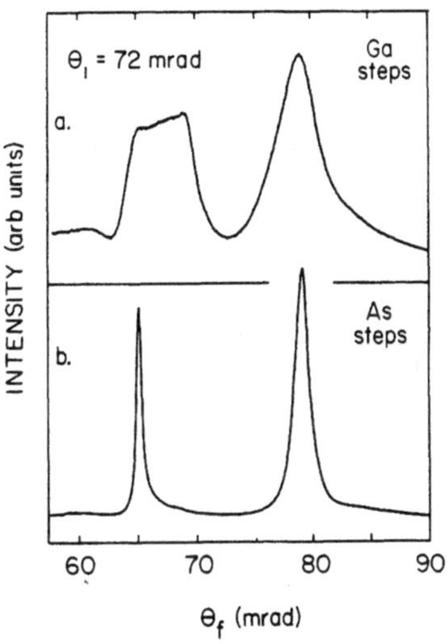

Fig. 11. Specular beam profile after growth is stopped for GaAs(100) misoriented by 35 mrad toward (111)A (top curve) and (111)B (bottom curve) planes. The incident angle is 72 mrad and the beam is directed down the staircase. The broader split peak in the top curve indicates more disorder in the terrace distribution on substrates misoriented toward (111)A.

It is difficult to distinguish what may be an anisotropic diffusion coefficient from differences in reflectivities at steps. This is understood when we look at the amount of disorder within the terrace distributions on the steps disorder when growth is stopped. The As terminated steps remain ordered. Other measurements show that the Ga terminated step edges are less kinked than the As steps, implying a possibly lower reactivity at the Ga edges.

4.3 Ge Growth

Many of the same growth mechanisms that are observed on GaAs are also observed on Ge. What makes this system perhaps more amenable to analysis is that the adatom species is a monomer and the reconstruction is stable over a wide temperature range. Figure 12 shows Ge RHEED oscillations on a 1 mrad and a 7 mrad surface. Aside from the bilayer period which will be discussed shortly, one should note that the high density of steps on the 7 mrad misorientation diminishes cluster formation on the terraces, much like that on GaAs.

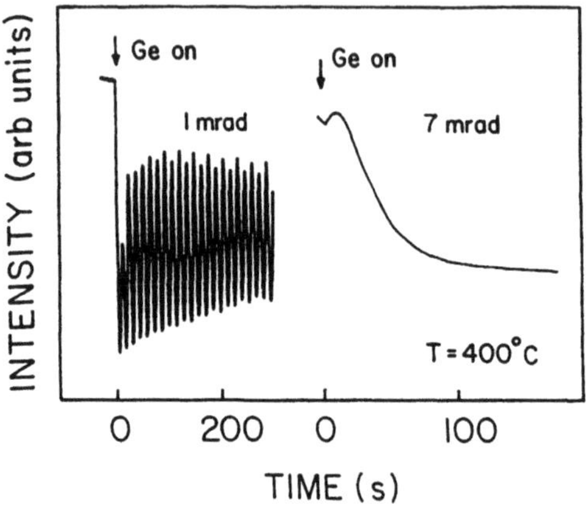

Fig. 12. RHEED oscillations observed during the growth of Ge. Strong oscillations are observed when the surface is within 1 mrad of the (100). No oscillations are observed when the surface is misoriented by 7 mrad toward the <010> under the same growth conditions. Incident angle is 65 mrad and the beam is directed in a <010> azimuth.

The extinction point of these oscillations was measured as a function of growth rate. Under constant background conditions this occurred at a fixed temperature independent of growth rate [33]. However with the fluctuation of residual As levels (which was the main contaminant) the extinction temperature varied. The trend was to higher temperatures with higher background levels. Figure 13 shows RHEED oscillations with unintentional and intentional As flux applied. Clearly the addition of As enhances clustering.

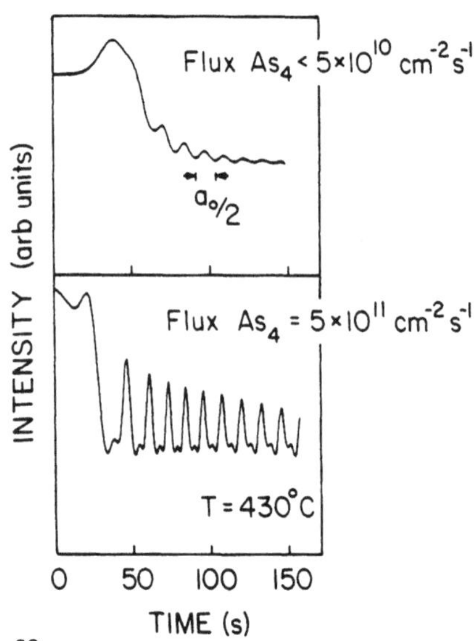

Fig. 13. Ge RHEED oscillations with and without impurity flux. Stronger oscillations are observed in the presence of As^4 (lower curve). This indicates greater clustering with impurities. Growth is by monolayers of Ge but repeat period of oscillations corresponds to deposition of two monolayers. The incident angle is 20 mrad and the beam is directed 15° from a <011> azimuth.

The cluster formation shows up dramatically in the RHEED pattern. If the electron beam is directed along a [010] axis, diverging streaks are observed to emanate from each of the diffracted beams. Diffracted beam profiles across these streaks are shown in Fig.14. This plot should be compared against that of Fig.7. The separation in $\Delta\phi_f$ is plotted as a function of final angle in Fig.15 along with that expected from Fig.7. Clearly the effect of clusters with straight edges along <011> directions is being observed. The diffracted behavior is simply a geometric shape effect as described in Sec.3.2.

The fact that the superstructure 2×1 relections show broadening along only one axis indicates that the clusters are rectangular [31]. The long axis of the cluster corresponds to the non-reconstructed direction. The superposition of the two possible domains makes the diffraction pattern symmetric about $\Delta\phi_f$=0 for

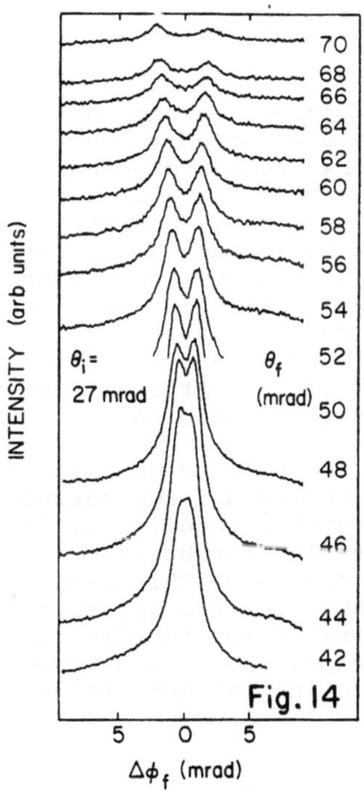

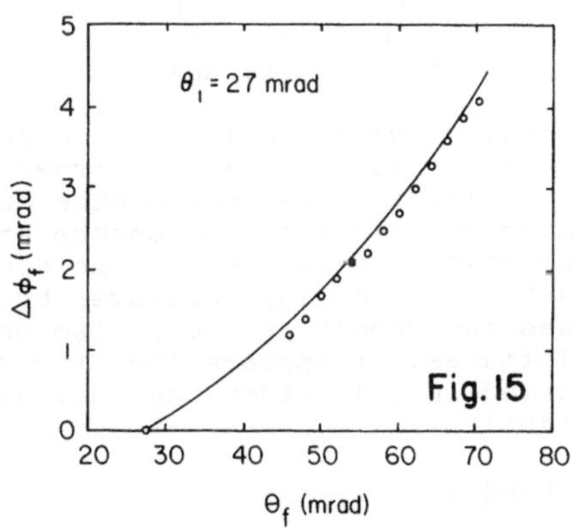

Fig. 14. Intersecting diffraction streaks along θ_f accompany Ge growth oscillations. Diffracted profiles are taken at various final angles perpendicular to the streak. Incident angle is 27 mrad and beam is directed along a <010> azimuth.

Fig. 15. Separation between streaks in Fig.14 as a function of final angle. Circles are data and solid line is calculated assuming clusters are rectangular with edges oriented along <011> directions.

the integral order beams. The possibility that anti-phase disorder in the 2×1 reconstruction contributes to the streaking is ruled out because the maximum intensity is located close to the integral order beams.

The shape of the clusters is likely governed by energetics. Having edges along <011> directions minimizes the number of dangling bonds for a given area cluster [4]. More difficult to explain is the elongation of the clusters. A monolayer high cluster that is square contains two types of step edges. These are distinguished by whether the dangling bond is pointed along or perpendicular to the step edge. Arsenic is known to cause a mass migration of atoms when applied to a stepped surface of Ge [32]. This was attributed to the As occupying (111) sites at only one of the types of step edges and pinning its motion. Anisotropic diffusion could also play a role in the elongation [6].

The clusters are stable when growth is stopped. The streaking only disappears when the substrate temperature is increased to a level where normally no oscillations are observed (>550°C). Presumably, at this point the dissociation rate of critical clusters greatly exceeds their formation rate.

To understand the role of anisotropic growth in more detail, the growth of Ge was followed on a surface misoriented by 4 mrad toward a <011> direction. The high temperature non-growth surface contained a two domain, $2\times1 + 1\times2$, diffraction pattern. Each domain is due to an alternate sublattice of the diamond crystal structure. Growth conditions were chosen above the temperature where oscillations were observed [34]. Under these conditions, the growth caused the 2×1 domain to predominate. This can be explained by mechanisms similar to those outlined for the asymmetric cluster formation. The kinetics can be such that either anisotropic diffusion or different reflectivities of adatoms at steps can cause one step to propagate at the expense of the other.

Finally, we would like to comment on the bilayer oscillations. Sakamoto et al [35] have observed these on the Si(100) surface, and we find the same dependence on diffraction conditions. The bilayer oscillations in essence are directly attributable to the alternating sublattices of the diamond structure. This means that growth is proceeding monolayer by monolayer but the diffraction senses the repeat period of two monolayers or the two distinct sublattices. It appears that the anisotropic growth processes observed in this study can only help make the two sublattices more distinct.

5. Summary

The fundamental mechanisms involved in crystal growth by MBE were outlined. Included in a modified BCF model were the effects of adatom and impurity adsorption, reconstruction, surface diffusion, cluster formation, and step propagation. RHEED was then used to probe the surface reconstruction and the step structure during the growth of GaAs and Ge. For both systems the competing processes of cluster formation and step propagation were followed as a function of growth conditions. At low temperatures or in the presence of impurities, layer-by-layer clustering is strongly enhanced. Here, strong RHEED intensity oscillations are observed. For Ge, strong cluster shape effects are observed in the diffraction. By introducing an array of steps due to a misorientation from a low-index plane, step propagation becomes more important and the oscillations diminish. For both surfaces, strong anisotropic growth effects were observed as the azimuthal orientation of the misorientation was changed. The issues that remain unresolved include: (1) separating diffusion processes from the kinetics of cluster formation/dissociation, (2) understanding the contributions to anisotropic step growth growth more completely, and (3) incorporating the role of non-equilibrium diffusion processes.

ACKNOWLEDGEMENTS

This work was supported in part by the Ceramics and Electronics Materials program of the National Science Foundation under Grant No. DMR-8615207.

REFERENCES

[1] J.R. Arthur, Jr., J. Phys. Chem. Solids (1968) 39, 4032.
[2] C.T. Foxon and B.A. Joyce, Surface Science (1975) 50, 434.
[3] W.K. Burton, N. Cabrera, and F.C. Frank, Phil. Trans. Roy. Soc. (1951) 243, 299.
[4] L.J. Giling and W.J.P Enckevort, Surface Science (1985) 161, 567.
[5] J. Van Vechten, J. Crystal Growth (1977) 38, 139.
[6] A. Rockett, Eighth MBE Conference, Los Angeles, 1987.
[7] J.M. Van Hove, P.I. Cohen, and C.S. Lent, J. Vac. Sci. Technol. (1983) A1, 546.
[8] A. Madhukar, Surface Science (1983) 132, 344.
[9] K. Voigtlaender, H. Risken, and E. Kasper, Appl. Phys. (1986) A39, 31.
[10] V. Fuenzilada and I. Einsele, J. Crystal Growth (1986) 74, 596.
[11] R. Ghez, S.S. Iyer, and S.L. Delage, J. Vac. Sci. Technol. (1987) B5, 740.
[12] G.S. Petrich, G.J. Whaley, P.R. Pukite, and P.I. Cohen, unpublished.
[13] F. Reif, Fundamentals of Statistical and Thermal Physics, (McGraw-Hill, NY, 1965).
[14] D. Walton, J. Chem. Phys. (1962) 37, 2182.
[15] G. Gilmer and P. Bennema, Crystal Growth, P.Hartman, ed., (North-Holland, Amsterdam, 1973); P.I.Cohen, unpublished.
[16] P.R. Pukite, PhD Thesis, University of Minnesota, unpublished.
[17] P.R. Pukite, J.M. Van Hove, and P.I. Cohen, Appl. Phys. Lett. (1984) 44, 456.
[18] P.R. Pukite, C.S. Lent, and P.I. Cohen, Surface Science (1985) 161, 39.
[19] P.R. Pukite, J.M. Van Hove, and P.I. Cohen, J. Vac. Sci. Technol. (1984) B2, 243.
[20] C.S. Lent and P.I. Cohen, Surface Science (1984) 139, 121; C.S. Lent and P.I. Cohen, Phys. Rev. (1986) B33, 8329.
[21] J.H. Neave, B.A. Joyce, P.J. Dobson, and N. Norton, Appl. Phys. (1983) A31, 1.
[22] J.M. Van Hove, C.S. Lent, P.R. Pukite, and P.I. Cohen, J. Vac. Sci. Technol. (1983) B1, 741.

[23] B.F. Lewis, F.J. Grunthaner, A. Madhukar, T.C. Lee, and R. Fernandez, J. Vac. Sci. Technol. (1985) B3, 1317.
[24] E. Bauer, RHEED and Reflection Electron Imaging of Surfaces, ed. by P.K. Larsen, (Plenum Press, NY, in press).
[25] S.T. Purcell, B. Heinrich, and A.S. Arrott, Phys. Rev. (1987) B35, 6458.
[26] P.I. Cohen, P.R. Pukite, J.M. Van Hove, and C.S. Lent, J. Vac. Sci. Technol. (1986) A4, 1251.
[27] J.M. Van Hove, P.R. Pukite, and P.I. Cohen, J. Vac. Sci. Technol. (1985) B3, 563.
[28] J.M. Van Hove and P.I. Cohen, J. Crystal Growth, (1987) 50 .
[29] J.M. Van Hove and P.I. Cohen, Appl. Phys. Lett. (1986) 47, 726.
[30] J.H. Neave, P.J.Dobson, B.A. Joyce, and J. Zhang, Appl. Phys. Lett. (1985) 47, 100.
[31] J.A. Martin, D.E. Savage, W. Moritz, and M.G. Lagally, Phys. Rev. Lett. (1986) 56, 1936.
[32] P.R. Pukite and P.I. Cohen, J. Cryst. Growth (1987) 50, 1739.
[33] J. Aarts, W.M. Gerits, and P.K. Larsen, Appl. Phys. Lett. (1986) 48, 931.
[34] P.R. Pukite, S. Batra, and P.I. Cohen, RHEED and Reflection Electron Imaging of Surfaces, ed. by P.K. Larsen, (Plenum Press, NY, in press).
[35] T. Sakamoto, N.J. Kawai, T. Nakagawa, K. Ohta, and T. Kojima, Appl. Phys. Lett. (1985) 47, 617.
[36] H.J. Leamy and G.H. Gilmer, J. Crystal Growth (1979) 24/25, 499.
[37] R. Gomer, in Surface Mobilities on Solid Marerials, ed. by V. T. Bien, (Plenum Press, NY, 1983) p.15

Laser-Induced Desorption Determinations of Surface Diffusion on Rh(111)

E.G. Seebauer[1]* *and L.D. Schmidt*[2]

[1]Laser and Atomic Physics Divison,
Sandia National Laboratories, Albuquerque,
New Mexico 87185, USA

[2]Department of Chemical Engineering and Materials Science,
University of Illinois, Urbana, IL 61801, USA

Surface diffusion of hydrogen, deuterium and CO on Rh(111) has been investigated by laser-induced thermal desorption (LITD) and compared with previous results for these species on Pt(111) and on other metals. For deuterium in the coverage range $0.02 < \theta < 0.33$, the pre-exponential factor $D_0 = 8 \times 10^{-2}$ cm^2/s, with a diffusion activation energy $3.7 \leq E_{diff} \leq 4.3$ kcal/mol. For CO, $E_{diff} = 7$ kcal/mol, but D_0 rises from 10^{-3} to 10^{-2} cm^2/s between $\theta = 0.01$ and 0.40.

Values of E_{diff} on different surfaces appear to correlate with differences in heats of adsorption in different binding states which form saddle point configurations in surface diffusion. In addition, oxidation reactions on Rh and on several other transition metal surfaces may be limited to CO or H surface diffusion.

1. INTRODUCTION

Surface diffusion of adsorbates on transition metals has proven very difficult to characterize, and only a few systems have been studied [1-11]. Recently, the technique of laser-induced thermal desorption (LITD) has been used for such experiments [4,5]. This method exploits the ability of a pulsed, focused laser beam to desorb species from a small, well-defined area while leaving the rest of the surface unperturbed. The desorbed molecules are detected as a background pressure rise in the ultrahigh vacuum chamber. The laser beam is then blocked, material adsorbed outside the spot allowed to diffuse in, and after a suitable period of time, the adsorbate which has diffused into the spot is detected by a second laser pulse. Fick's second law for transient diffusion is then applied to calculate E_{diff} and D_0 for diffusion [5].

We have employed LITD to investigate the surface diffusion of H and D atoms and CO molecules chemisorbed on a Rh(111) single crystal. We compare our results with the growing body of diffusion data that has appeared in the literature in recent years and identify important trends.

2. EXPERIMENTAL

The ultrahigh vacuum chamber has been described previously [5]. The Rh(111) sample was a disk 1 mm thick and 6 mm in diameter, oriented to within 0.5° of (111), and polished by standard metallographic techniques. After cleaning, TPD spectra of CO and hydrogen identical to those reported in the literature [13] were obtained.

---

*Present address: Department of Chemical Engineering,
University of Illinois, Urbana, IL 61801

Light from the 9.6μ P(20) line of a Lumonics TEA CO_2 laser was apertured and focused onto the crystal to give a 1.2 x 0.4 mm rectangle. Mass spectrometer desorption pulses were recorded and stored using a digital multichannel analyzer.

3. RESULTS

Diffusion data were modeled using an expression derived from Fick's second law with the assumptions of a constant diffusion coefficient and an initial step concentration profile, as described previously [5]. Arrhenius plots for the diffusivity of deuterium determined this way are shown in Fig. 1 for several initial coverages. Values for E_{diff} and D_0 extracted from plots like these are shown in Fig. 2. As θ varies from 0.02 to 0.33, E_{diff} rises slightly from 3.7 to 4.3 ± 0.4 kcal/mol while D_0 remains near 8 x 10^{-4} cm^2/s. Thus D_0 is within a factor of 10 of the "normal" value of 2.5 x 10^{-3} cm^2/s [14].

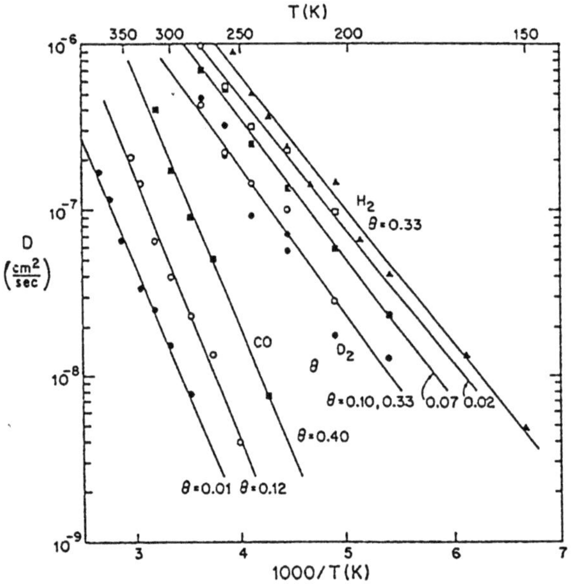

Fig. 1. Arrhenius plots of the diffusivity D of hydrogen, deuterium and CO on Rh(111) at initial coverages indicated

Figure 1 also includes an Arrhenius plot for hydrogen diffusion at θ = 0.33. The value (8 x 10^{-4} cm^2/s) for D_0 matches that for deuterium, but E_{diff} = 3.7 kcal/mol for hydrogen is 0.6 kcal/mol lower than for deuterium.

Arrhenius plots for CO diffusion are shown in Fig. 1 as well, with the diffusion parameters plotted in Fig. 2. As θ increases from 0.01 to 0.40, E_{diff} remains essentially constant at 7.1 ± 0.5 kcal/mol, but D_0 rises from 2 x 10^{-3} to 3 x 10^{-2} cm^2/s. Thus, the rate of diffusion increases with θ, in contrast to the case of deuterium (and presumably hydrogen) where the diffusion rate decreases.

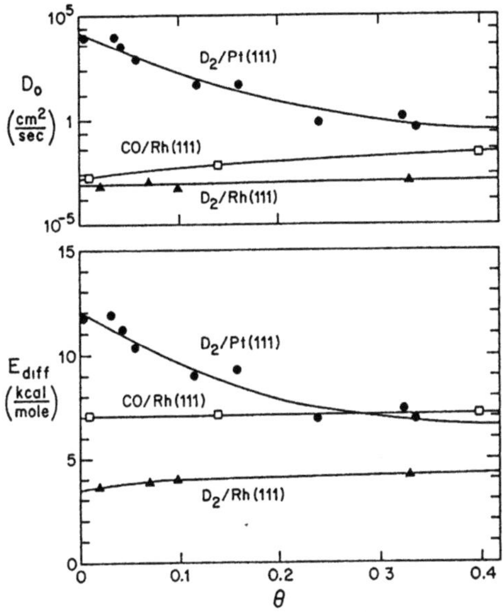

Fig. 2. Plots of E_{diff} and D_0 as a function of θ for hydrogen and CO on Rh(111) and Pt(111)

4. DISCUSSION

4.1 Comparison Among Metals

There is now sufficient surface diffusion data for gases adsorbed on metals to make useful comparisons possible. Figure 3 shows Arrhenius plots for several systems [3-5, 7-9, 11, 12] from which it is evident that values of D at a particular temperature differ by more than 10^3 and that E_{diff} and D_0 values vary considerably.

Typically for hydrogen, E_{diff} lies between 4 and 6 kcal/mol, regardless of the substrate metal. Since the activation energies E_d for atomic (not recombinatory) desorption usually are near 60 kcal/mol on transition metals, the ratio of E_{diff} to E_d is ~0.06 to 0.1 for H on all metals. In situations where coverage variations were examined (Pt, Rh, W), E_{diff} often rises slightly (by 0.5 to 1 kcal/mol) as θ increases. D_0 is far less predictable with values ranging from 1×10^{-7} to 3×10^4 cm^2/s.

The only other atomic systems for comparison are O and N on W [1,2,9,10,15,16,17,50,51]. These adsorbates are much more strongly bound to the surface, so that E_{diff} is considerably higher (usually ~25 kcal/mol), with E_{diff}/E_d ~ 0.15 to 0.2.

Measurements of molecular surface diffusion appear to be limited to CO. Very fast diffusion has been measured by LITD on Ni(100) [7] with E_{diff} = 5 kcal/mol and D_0 = 0.2 cm^2/s. An indirect measurement of CO diffusion on Pt(111) has been performed with He beam scattering [6], yielding E_{diff} = 7 kcal/mol, similar to that on Rh(111). Thus, E_{diff} is about 0.23. LITD experiments in our laboratory [5] established an upper bound for D_0 of ~10^{-4} cm^2/s if E_{diff} = 7 kcal/mol is assumed.

That D_0 for CO/Pt(111) is a factor of 100 smaller than for Rh(111) is quite surprising since previous CO adsorption studies have indicated that

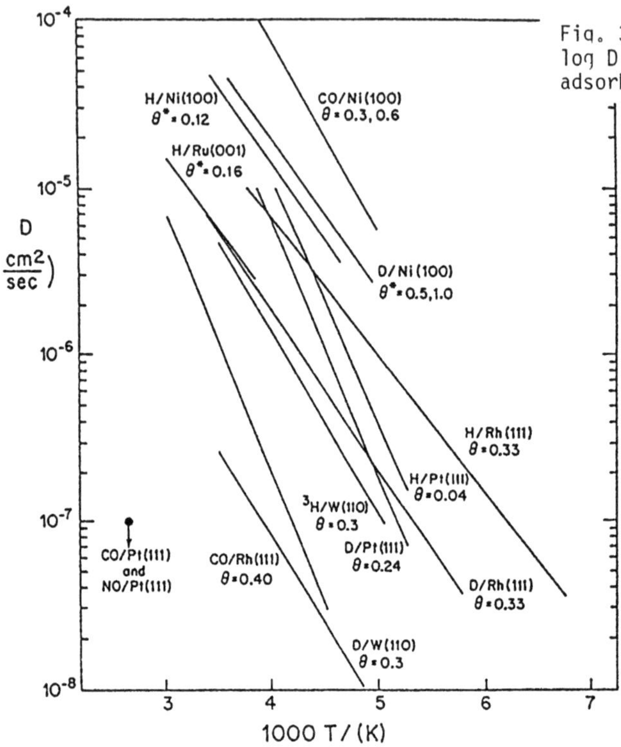

Fig. 3. Arrhenius plots of log D vs. 1/T for several adsorbate systems

the two metals behave quite similarly. However, desorption studies examine mainly the potentials normal to the surface. The potentials parallel to the surface are evidently quite different.

4.2 Potential Energy Surfaces

Surface diffusion involves the motion of a chemisorbed atom or molecule from one adsorption site, over a saddle point to another site. In this section, we consider the possible saddle point configurations and the relationships between diffusion and desorption.

In Table 1 are shown E_d values calculated for each binding state of several adsorption systems for which E_{diff} is known. Multiple binding states of H_2 on Rh(111) are not well-resolved and the E_d values shown are at low and high coverage. Note that for a dissociated species the bond energy of an H atom $E_{dH} = 1/2(E_{dH_2} + E_{H-H})$. The activation energy for surface diffusion should be at least the difference between the heats of adsorption in strongly and weakly bound states because the saddle point configuration for migrations of the tightly-bound state should approximate the bonding configuration of the weaker site. The barrier to surface diffusion could, of course, be greater than the energy difference, particularly if the weaker state is still fairly strongly bound and, therefore, represents a deep local minimum in the potential.

It is seen from Table 1 that the difference in E_d between different binding states corresponds fairly well with E_{diff} for both molecular CO and

TABLE 1

Diffusion and Desorption Activation Energies

Surface	CO E_d	CO E_{diff}	CO ΔE_d	H_2 or D_2 E_{dH_2}	E_{dH}	E_{diff}	ΔE_{dH_2}	ΔE_{dH}
Rh(111)	31,26 [13]	7.1	5	19,17 [13]	62 5,61 5	3.7-4.3	2	1
Pt(111)	30,22 [25]	≥7 [6,12]	8	18,13,8 [25]	60,57 5,55	12-7 [5]	5,10	2 5,5
Ni(100)	13,18,22,28 5 [27]	5 0 [7]	6.5,10 5, 15.5	12,20 6 [26]	58,62 5	4.0 [4,7]	8 6	4 3
Ru(001)	--	--	--	22,29 [29]	63,66 5	4 0 [8]	7	3 5
W(110)	--	--	--	27,33 [30]	65 5,68 5	4-5 [3,9,11]	6	3

All activation energies in kcal/mole References are indicated in square brackets.

the atomic adsorbates H and H, suggesting that the diffusion barrier may indeed be the difference ΔE_d. Although E_{diff} is sometimes larger than ΔE_d, particularly for hydrogen, rarely is E_{diff} smaller than ΔE_d. Only N on W(110) is a significant exception, where E_{diff} is 25% less than ΔE_d. In this case, even the weaker state has a high binding energy of 125 kcal/mol. The energy at the saddle point between the two stable states may thus be quite large, so that diffusion from one strongly bound state to another may proceed through another saddle altogether.

4.3 Implications for Surface Reactions

The condition $E_R \geq E_{diff}$ must hold for any surface reaction between two adsorbed species [15] that proceeds by a Langmuir Hinshclwood (I-H) mechanism. CO oxidation is the L-H reaction that has been studied in the most detail. E_R is generally within a few kcal/mol of 7 kcal/mol [17,18] which is E_{diff} for CO on Rh(111) and Pt(111). Assuming that E_{diff} lies near 7 kcal/mol for CO on other metals and crystal planes, it appears likely that surface diffusion of CO is indeed the rate limiting step for CO oxidation on Rh(111) [18], Rh(poly) [19], Ir(111) [20], Pt(poly) [21], Pd(poly) [22], Pt(111) [17], and possibly Pd(111) [23].

However, E_R = 5.3 kcal/mol for hydrogen oxidation on Rh(111) [24], nearly matching E_{diff} = 4 kcal/mol for hydrogen diffusion. Thus, the rate limiting step may be hydrogen diffusion.

5. SUMMARY

The present results show that surface diffusion coefficients can vary by several orders of magnitude for adsorption systems with nearly identical heats of adsorption. While surface diffusion probably controls many surface reaction processes, surface diffusion coefficients are very difficult to measure and to predict using any available theories or empirical correlations. Sufficient data are now being accumulated so that it may soon be possible to compare potential surfaces predicted theoretically with E_{diff} and to describe pre-exponential factors in terms of barrier crossing events.

ACKNOWLEDGMENT

This work was performed at the University of Minnesota and supported by the U.S. Department of Energy under contract number DE-AC04-76DP00789 and by NSF under Grant No. DMR 82126729.

REFERENCES

1. A. Polak and G. Ehrlich, J. Vac. Sci. Technol. 14, 407 (1977)
2. R. Butz and H. Wagner, Surface Sci. 63, 448 (1977)
3. R. DiFoggio and R. Gomer, Phys. Rev. B25, 3490 (1982)
4. S. M. George, A. M. DeSantolo and R. B. Hall, Surface Sci. 159, L425 (1985)
5. E. G. Seebauer and L. D. Schmidt, Chem. Phys. Lett. 123, 129 (1986)
6. B. Poelsema, L. K. Verheij and G. Comsa, Phys. Rev. Lett. 49, 1731 (1982)
7. D. A. Mullins, B. Roop and J. M. White, Chem. Phys. Lett. 129, 511 (1986)
8. C. H. Mak, J. L. Brand, A. A. Deckert and S. M. George, J. Chem. Phys. 85, 1676 (1986)
9. M. Tringides and R. Gomer, Surface Sci. 155, 254 (1985)
10. M. Tringides and R. Gomer, J. Chem. Phys. 84, 4049 (1986)
11. S. C. Wang and R. Gomer, J. Chem. Phys. 83, 4193 (1985)
12. E. G. Seebauer, A. C. F. Kong and L. D. Schmidt, Surface Sci. 176, 134 (1986)
13. T. W. Root, L. D. Schmidt and G. B. Fisher, Surface Sci. 150, 173 (1985)
14. R. Gomer: In Surface Mobilities on Solid Materials, ed. by V. T. Binh (Plenum, NY, 1983)
15. G. Ehrlich and K. Stolt, Ann. Rev. Phys. Chem. 31, 603 (1980); G. Ehrlich and J. F. Hudda, J. Chem. Phys. 35, 1421 (1961)
16. D. L. Freeman and J. D. Doll, J. Chem. Phys. 78, 6002 (1983)
17. C. T. Campbell, G. Ertl, H. Kuipers and J. Segner, J. Chem. Phys. 5862 (1980); J. L. Gland and E. B. Kollin, J. Chem. Phys. 78, 963 (1983)
18. S. B. Schwartz, L. D. Schmidt and G. B. Fisher, J. Phys. Chem. 90, 6194 (1986)
19. C. T. Campbell, S.-K. Shi and J. M. White, Appl. Surface Sci. 2, 382 (1979)
20. P. A. Zhdan, G. K. Boreskov, W. F. Egelhoff, Jr. and W. H. Weinberg, Surface Sci. 61, 377 (1976); V. P. Ivanov, G. K. Boreskov, W. F. Egelhoff, Jr. and W. H. Weinberg, J. Catalysis 48, 269 (1977)
21. H. Heyne and F. C. Tompkins, Proc. Roy. Soc. A292, 460 (1966)
22. T. Matsushima and J. M. White, J. Catalysis 39, 265 (1975)
23. G. Ertl and M. Neumann, Z. Physik, Chem. NF 90, 127 (1974); T. Engel and G. Ertl, J. Chem. Phys. 69, 1267 (1978).
24. J. T. Yates, Jr., P. A. Thiel and W. H. Weinberg, Surface Sci. 82, 45 (1979)
25. R. W. McCabe and L. D. Schmidt, Surface Sci. 65, 189 (1977)
26. B. E. Koel, D. E. Peebles and J. M. White, Surface Sci. 125, 709 (1983)
27. M. Kiskinova and D. W. Goodman, Surface Sci. 108, 64 (1981)
28. P. W. Tamm and L. D. Schmidt, Surface Sci. 26, 286 (1971)
29. P. Fuelner and D. Menzel, Surface Sci. 154, 465 (1985)
30. P. W. Tamm and L. D. Schmidt, J. Chem. Phys. 54, 4775 (1971)

Mobility of Rare Gases Adsorbed on the (001) Surface of MgO

M. Karimi[1] *and G. Vidali*[2]

[1]Utica College, Physics Department, Utica, NY 13502, USA
[2]Syracuse University, Physics Department, Syracuse, NY 13244, USA

1. Introduction

Recently, there has been considerable interest in studying the adsorption of rare gases or simple molecules on the (001) surface of MgO[1,2]. The principal motivation of many experimental studies is the possibility of observing the effects of substrate geometry on structural and thermodynamic properties of the adlayers. There are at least two important effects to study.

First, there is the effect of a higher corrugation of the surface (on the atomic scale, as seen by the adatom) on the mobility of the adatoms as well as on the development of the various adlayer phases. It is particularly interesting and illuminating to make comparisons with a system that has been extensively studied, i.e., rare gases on graphite. In the latter case, the substrate holding potential is stronger than in the MgO case, yet the surface corrugation, as determined by He diffraction studies [3-6] is larger for MgO than for graphite. One of the purposes of this calculation is to quantify this difference for other rare gas atoms or molecules of interest and see the implications of such effects on surface mobilities and other properties of the adlayer [7].

Second, there is the effect of the substrate geometry on the development of the various phases of the adlayers. For example, in the case of graphite, which has a surface with triangular symmetry, commensurate as well as incommensurate adlayers can be seen, sometimes for the same gas adatom, depending on the subtle interplay of substrate geometry, corrugation, adlayer holding potential, adlayer-adlayer interactions and structure of the adlayer atoms in the bulk [8]. The picture that starts to emerge is one in which the substrate holding potential is not strong enough to force the rare gas atoms in registry with the substrate. On the contrary, so far only incommensurate phases of rare gases have been observed on MgO. Methane, on the other hand, grows with a square c(2x2) lattice on MgO, despite the fact that bulk methane has a hexagonal symmetry [9].

To shed some light on these effects and better understand the mechanisms of growth and transformation of adlayers on substrates, we have performed calculations to obtain the interaction potentials between He, Ar, Kr, Xe or CH_4 and the (001) surface of MgO. From these calculations we have obtained, among other quantities, the corrugation of the potential across the surface and the activation energy for surface diffusion of an adatom (Section 2). We have then compared these results with previous calculations done for the same adatoms on graphite; we will show (Section 3) that many experimentally observed differences of adlayer properties (commensurate vs. incommensurate lattices, adatom mobilities, etc.) can be explained by our calculations.

2. The Model

The model for the interaction potential that we are using has been described in detail in Refs. [10-17]; therefore only an outline will be given here. The potential can be written:

$$V(\vec{r}) = V_A(\vec{r}) + V_R(\vec{r}) - 1/2\ \alpha E^2(\vec{r}), \tag{1}$$

where α is the static polarizability of the adatom and $E(\vec{r})$ is the ionic electric field. The attraction part $V_A(r)$ is a superposition of damped dipole and damped quadrupole interactions [10-15]:

$$V_A(\vec{r}) = \sum_i V_i(|\vec{r} - \vec{r}_i|), \tag{2a}$$

$$V_i(x) = -\sum_{n=3}^{4} C_{2n}^{(i)} f_i(x) x^{-2n}, \tag{2b}$$

where i labels the ionic sites, x is the distance between the adatom and the crystal ions, and the damping functions $f_i(x)$ are those proposed by Tang and Toennies [18], which have been tested for gas phase atom-atom interactions and recently for He-LiF [10,11,12]

$$f_i(x) = 1 - \sum_{k=0}^{6} \frac{(\beta_i X)^K}{k!} \exp(-\beta_i x) \tag{3}$$

and β_i is the softness parameter of the charge density of the surface. C_6^- and C_6^+ were calculated using the London expression [19-21]:

$$C_6^{\pm} = \frac{3}{2} \alpha_{\pm} \alpha \frac{E_{\pm} E}{E_{\pm} + E}, \tag{4}$$

where $\alpha_{\pm}$ and $E_{\pm}$ are the ionic dipole polarizability and corrected ionization potentials of the ions, and α and E are the same variables for the adatom [22]. Values of $E_+ = 177.3$ ev, $E_- = 23.6$ ev, $\alpha_+ = 0.069 Å^3$, and $\alpha_- = 1.679 Å^3$ were taken from Refs. [23,24]. The expression for the parameters C_8^+ and C_8^- is given by [25]

$$C_8^{\pm} = \frac{9}{4} C_6^{\pm} \sqrt{a_0} \left(\left(\frac{\alpha}{z_0}\right)^{1/2} + \left(\frac{\alpha_{\pm}}{z_{\pm}}\right)^{1/2} \right), \tag{5}$$

where a_0 is the Bohr radius, z_0 and $z_{\pm}$ are the number of outer electrons of adatom and ions, respectively.

The repulsive part $V_R(\vec{r})$ has been evaluated using the effective medium theory which assumes a proportionality between $V_R(\vec{r})$ and the weighted average of the charge density ρ of the solid. Following Refs. [26,27], we have:

$$V_R(\vec{r}) = \alpha_{eff}\ \bar{\rho}(\vec{r}) \quad , \tag{6a}$$

$$\alpha_{eff} = \alpha_0 - \alpha_{at} \quad , \tag{6b}$$

$$\alpha_{at} = \int_\Omega \phi_a\ (\vec{r}'- \vec{r})d\vec{r}' \quad , \tag{6c}$$

$$\bar{\rho}(\vec{r}) = 1/\alpha_{at} \int_\Omega \phi_a\ (\vec{r}'- \vec{r})\rho(\vec{r}')d\vec{r}', \tag{6d}$$

where Ω is the volume of a sphere with a finite radius R_c [27]. The weighting function is the electrostatic potential ϕ_a of the adatom. This theory was applied successfully to rare gas atoms-insulator systems [11,28]. In the case of a He atom, the coefficient α_0 in [6] has been estimated to be about 42 ev - Å^3 [11,29]. Our charge density $\rho(\vec{r})$ is obtained by adding contributions from individual ions:

$$\rho(\vec{r}) = \sum_i \rho_i(\vec{r} - \vec{r}_i) \tag{7}$$

where $\rho_i(\vec{r} - \vec{r}_i)$ is taken to be the crystal charge density of the ith ion at position $\vec{r}_i$, [30]. ρ_i for each ion was fitted to a simple exponential $\rho_i(r) = \rho_{i0}e^{-\beta_i r}$. The parameters are $\rho_{-0} = 13.79$ Å^{-3}, $\beta_- = 3.525$ Å^{-1} and $\rho_{+0} = 42.10$ Å^{-3}, $\beta_+ = 5.70$ Å^{-1} for O^{2-} and Mg^{2+}, respectively.

The third term in Eq. (1) is the induced dipole energy. This has been evaluated many times in the literature [10]. The lateral average of this term is

$$-\frac{1}{2}\alpha E^2 = -\frac{64\pi^2\alpha e^2}{a^4\ (1 + \exp\ (-\pi\sqrt{2}))^2}\exp\left(\frac{-4\pi}{a}z\right) \quad , \tag{8}$$

where e is the charge of the electron, and a is the lattice constant.

The z-dependent Fourier components $V_G(z)$ of the atom-surface potential are:

$$V(r) = \sum V_G(z)e^{iG\cdot R} \quad , \tag{9a}$$

$$G = 2\pi/a\ (m_1,m_2) \quad , \tag{9b}$$

$$R = a(n_1,\ n_2) \quad , \tag{9c}$$

where m_1, m_2, n_1, n_2, are integers, a = 2.98Å for MgO surface unit cell, and $V_G(z)$ are computed using the formulas developed in Ref. [10] for an ionic fcc crystal:

$$V_G(z) = \frac{2\pi}{a^2}\sum_{m_1,m_2}\ \sum_{\ell,s} (-1)^{(\ell+s-1)\ (m_1 + m_2)} e^{iG\cdot R}\int_0^\infty tdtV_s(x)J_0(Gt) \tag{10}$$

where $\ell = 1$ for the first layer and $\ell = 2$ for the second layer etc, s = 1 and s = 2 are the values for Mg^{2+} and O^{2-}, respectively, and $V_s(x)$ is the pair potentials introduced in (2), (3), (6), (7), with $x = \sqrt{t^2 + z^2}$. For the Fourier components of the potential only the contribution of the top two layers were considered since other contributions are negligible [14]. The lateral averaged part of $V_G(z)$ reads [10-17]:

$$V_0(z) = \frac{2\pi}{a^2}\left[-(C_6^+ + C_6^-)\frac{\zeta(4,z/d)}{4d^4} - (C_8^+ + C_8^-)\frac{\zeta(6,z/d)}{6d^6} + \right.$$

$$\alpha_{eff}\left\{\frac{\bar{\rho}_+}{\rho_+}\,\frac{\rho_{+0}(1+\beta_+ z)\exp(-\beta_+ z)}{\beta_+^2} + \frac{\bar{\rho}_-}{\rho_-}\,\frac{\rho_{-0}(1+\beta_- z)\exp(-\beta_- z)}{\beta_-^2}\right\}$$

$$\left. - H_{00}^+(z) - H_{00}^-(z) - \frac{32\pi\alpha e^2 \exp([-4\pi/a]z)}{a^2(1+\exp(-\pi\sqrt{2}))^2}\right], \qquad (11)$$

where $d = a/\sqrt{2}$ is the spacing between the MgO layers, $\zeta(n,x)$ is the Riemann zeta function [31], $H_{00}(z)$ are due to damping and can be evaluated analytically

$$H_{00}^{\pm}(z) = \int_0^\infty [f^{\pm}(x) - 1]\left\{\frac{c_6^{\pm}}{x^6} + \frac{c_8^{\pm}}{x^8}\right\} t\,dt \qquad (12)$$

The Fourier components V_{G1}, V_{G2}, V_{G3} and V_{G4} and higher order ones are given in Refs. [10], [14], [16].

For He-Ar/MgO systems α_0 are fixed to 42 ev-Å^3 and 247 ev-Å^3 respectively [32] and then c_6^- is changed until the required fit to the experimental eigenenergies is achieved (c_6^+ and c_8^+ have minor roles in determining V(z)). For He interacting with MgO $C_6^- = 9738$ mev-Å^6 (obtained by fitting) while the calculated $C_6^- = 6600$ meVA^3. This discrepancy is related to the poor understanding of the long range interaction between the atom and the ion in the solid. For other systems we multiplied the calculated value of C_6^+ with the factor C_6^-(fit)/C_6^-(calc) =1.17 for argon and α is changed until the experimental data (i.e. binding energies) are obtained.

When two adatoms interact in the vicinity of the surface, their gas phase interaction is modified due to the electromagnetic screening of the Van der Waals interaction by the substrate [33,34]. McLachlan has computed the following expression for the change of the adatom-adatom interaction [35].

$$\Delta U(r) = \frac{4c_{s1}}{r^6 p^{3/2}}\left[\frac{1}{3} - \frac{L^2}{pr^2}\right] - \frac{c_{s2}}{r^6 p^3}, \qquad (13)$$

where L is the height of the adatoms above an effective image plane (we adopt the relative L = <z> - d/2, where <z> is the expectation valueof the position of the atom is the laterally averaged part of the atom- surface potential), $p = 1 + 4L^2/r^2$, and r is their separation. Two simple formulas were developed in Ref. [34] for c_{s1} and c_{s2}:

$$c_{s1} = \frac{3}{4} g_0 \alpha_0^2 E_a x(2 + x)/(1 + x)^2 ,$$

$$c_{s2} = \frac{3}{4} g_0^2 \alpha_0^2 E_a x(x^2 + 3x + 1)/(1 + x)^3 ,$$

where $x = \frac{E_s}{E_a}$ and coefficients α_0, g_0, E_s, E_a for CH_4/MgO are taken from Ref. [33].

3. Results and Discussion

The He-MgO potential is reported in Figure 1; the minimum of the potential is at the top of the Mg ion [1]. From Fig. 1 one can calculate the corrugation of the potential going from the most to the least adsorptive site, as seen by an incoming He atom of 10 meV incident energy. Our result of 0.36 Å for the peak to trough corrugation is in very good agreement with atom beam scattering data (0.36 Å, see Ref. [4]; Ref. [5] has 0.46Å; Ref. [6] has 0.47 Å). Bound states energy levels are also in agreement with the measured ones [5], see Table I. As pointed out in the preceeding section, the coefficient c_6^- was allowed to change in order to fit

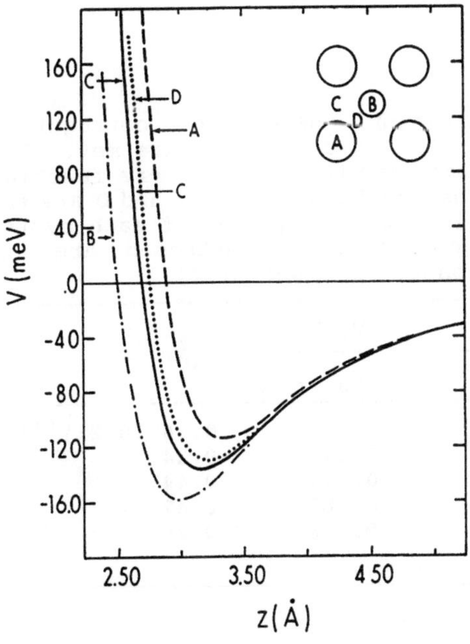

Fig. 1. Potential computed for a He atom above symmetry positions on an MgO (001) surface; shown at upper right: A(---), above anion; B(-.-), above cation; C(——), between anions; D(....), between anion and cation. The perpendicular distance is z.

Table I. Results of the potential energy calculations for gases above MgO; energy units are meV. $V_0(z_0)$ is the minimum of the laterally averaged potential, which occurs at z_0. α_0, c_6^- and c_6^+ are the potential parameters. E_0, E_1, E_2, and E_3 are the eigenvalues of $V_0(z)$, and <z> is the expectation value in the ground state.

atom	z_0(Å)	$\|V_0(z_0)\|$	α_0(eV-Å^3)	c_6^+(eV-Å^6)	c_6^-(eV-Å^6)	<z>
He	3.18	13.30	42	0.505	9.738	3.42
Ar	3.21	75.06	247	2.910	50.866	3.22
Kr	3.29	96.51	475	3.890	71.428	3.30
Xe	3.45	121.94	894	5.506	105.48	3.47
CH_4	3.26	99.10	296	3.790	70.733	3.29

$\|E_0\|$	$\|E_1\|$	$\|E_2\|$	$\|E_3\|$
10.25	5.67	2.82	1.25
72.53	67.63	62.96	58.52
94.57	90.78	87.08	83.50
120.2	116.81	110.30	104.00
94.50	85.81	77.70	70.14

the bound states data. This amounted to a 43% and 17% increase for He and Ar respectively. For Kr, Xe, and CH_4 a 17% increase correction was assumed. Similiar corrections had to be done for other systems [10]; it reflects the fact that C_6 coefficients are not precisely known, especially for ions in a solid, and that there are uncertainties in evaluating contributions from higher multiple terms.

Table II. Results of activation energy and corrugation calculations for gases above MgO; energy units are meV. The potentials V_A, V_B, V_C, and V_D are each the minumum, as a function of z, of V(r) at fixed x,y. The positions A, B, C, and D are the sites shown in Figure 1. ξ is the corrugation of each system evaluated at 10 meV. Values in the parentheses are the corresponding values for graphite from Ref. [17].

atom	$\frac{(V_A-V_B)}{\|V_0(z_0)\|}$	$\frac{(V_A-V_C)}{\|V_0(z_0)\|}$	$\frac{(V_A-V_D)}{\|V_0(z_0)\|}$	ξ(Å)
He	0.339 (0.225)	0.178	0.207	0.36 (0.21)[a]
Ar	0.265 (0.109)	0.108	0.140	0.24
Kr	0.345 (0.088)	0.191	0.217	0.44
Xe	0.303 (0.066)	0.166	0.187	0.42
CH_4	0.259	0.109	0.138	0.26

(a) From Ref. [3].

The fact that we obtain the correct corrugation and bound states for He-MgO suggests that our procedure has some validity. In Table I we have also reported the values of the well depth as well as the bound states for each system; in Table II the values of the difference in energies between different adsorption sites and corrugations for different systems are given. Results of the potential energy between like adatoms or molecules adsorbed on MgO for the noble gases He, Ar, Kr, Xe, and for the molecule CH_4 are presented in Table III. In Fig. 2 we have drawn the laterally averaged adatom-surface potentials.

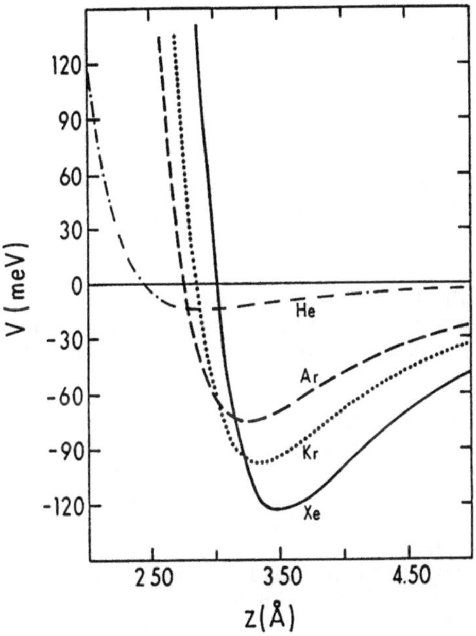

Fig. 2. Laterally averaged potential $V_0(z)$ for noble gases on (001) MgO.

We would like to make the following comments. For He and Ar, the corrugation of the potential as seen by the incoming atom decreases as the size of the atom increases; in particular the decrease in corrugation from He to Ar is compatible to what is observed for Ar/Ni and Ar/LiF scattering data [38,39]. The ratio (energy barrier)/(well depth) decreases with the size of the atom. This ratio is 0.34 for He/MgO while 0.22 for He/graphite (we are taking the maximum excursion in energy, i.e. from the least to the most favorable adsorption site). For Xe/MgO and Xe/graphite we have 0.30 and 0.07 respectively. Therefore it is relatively easier for an adatom to hop sites in the case of graphite than MgO. The more localized nature of adatom/MgO is substantiated by experimental data on mobility of CH_4 on MgO and graphite [2] (see below). In the case of CH_4/MgO the situation is more complicated. We are aware of only one experimental determination of the ground state energy of CH_4/MgO system; actually what is given is the isosteric heat of adsorption [2]. In absence of more information, and making reasonable assumptions about the temperature at which it was determined, we come up with a

Table III. Input parameters for the free space potential $U_0(r)$ and $\Delta U(r)$. D_0 and r_0 are the well depth and the location of the potential minimum for $U_0(r)$. D is the well depth for $U_0(r) + \Delta U(r)$. All lengths are in Angstrom units. D_0 and D are in meV.

atom	c_{s1} [(a)]	c_{s2} [(a)]	L	r_0	$\vert D_0 \vert$	$\vert D \vert$	$\Delta U(r)$
He	0.288	0.110	2.042	2.91 [(b)]	0.868 [(b)]	0.813	8.70%
Ar	15.175	6.155	2.076	3.76 [(c)]	12.22 [(c)]	11.149	9.61%
Kr	31.826	13.180	2.156	4.01 [(c)]	17.36 [(c)]	15.754	10.21%
Xe	92.492	37.485	2.316	4.36 [(c)]	24.17 [(c)]	21.261	13.68%
CH_4	28.513	11.419	2.133	3.88 [(c)]	18.66 [(c)]	16.974	9.93%

(a) c_{s1}, c_{s2} in units of eV-Å^6; noble gas values from Ref. [36]; parameters in c_{s1} and c_{s2} for CH_4/MgO system are from Ref. [33]
(b) r_0, D_0 values for He-He pair potential from Ref. [34].
(c) r_0, D_0 values for other systems from Ref. [35].

binding energy that is a bit high, between the Kr and Xe case. Now for both BN and graphite, experimentally determined binding energies of CH_4 and Kr are very close to each other (within a few percent).

We therefore decided to assume that the binding energy of CH_4 on MgO is the same as for Kr on MgO. We then built the potential and obtained $E_{ab} = 0.25$, see Table II. We don't have a value for CH_4/graphite but we estimate it, by comparison with other systems, to be correspondingly lower. This substantiates the previous qualitative discussion.
From Ref. [2] we have that the translational diffusive coefficient for CH_4/MgO is: $D \simeq 5 - 6\ 10^{-6}$ cm^2 sec^{-1} at 88 K and $\simeq 12\ 10^{-6}$ cm^2 sec^{-1} at 97 K. We write: [40]

$$D = D_0\ e^{-\Delta E / kT}$$

where ΔE is the activation energy for going from site to site; $D_0 = (1/4) d^2 \nu$, where d is the distance between two adsorption sites (d = 4.21 Å for CH_4 c(2x2) on MgO) and ν an "attempt frequency to overcome the barrier". We can crudely estimate ν in the following way. At the bottom of the well we can write $V(z) \simeq (1/2) k (z - z_{min})^2 - D$, where k is the "stiffness" of a classical harmonic oscillator (CH_4) in a well. We have:

$$\nu = \frac{\omega}{2\pi} = \sqrt{\frac{k}{m}}$$

where m is the reduced mass (we have assumed that CH_4 is a rigid molecule that interacts with 4 oxygen and a magnesium atom). We get: $D_0 \simeq 8\ 10^{-4}$ cm^2/sec. Using $E \simeq 24$ meV, $T \simeq 88$ K, we get $D \simeq 28\ 10^{-6}$ cm^2/sec, in fair agreement with the experimental value, $\simeq$ 5-6 10^{-6} cm^2/sec [2].

If we take the ratio of two experimentally determined diffusion coefficients, we get:

$$D_{88}/D_{97} = e^{-\frac{\Delta E}{k}\left(\frac{1}{88} - \frac{1}{97}\right)} \simeq 0.45,$$

which gives $\Delta E \simeq 60$ meV, while our calculation gives $\simeq 24$ meV. While this crude analysis gives the right order of magnitude, more puzzles remain to be solved (such as the factor 10 increase in viscosity of the CH_4 overlayer going from MgO to graphite [2]). We are presently working on these issues.

The effect of substrate screening is to primarily affect the strength of the adatom-adatom interaction, reducing the well depth of the potential by 8 - 15% from the free space values. Therefore, it should be incorporated into the calculations of a thin film adsorbed on a substrate.

We would like to point out that the experimentally determined corrugation for He/MgO is close to our value. A calculation done for He/LiF or NaCl using a similar method [11] gave a corrugation almost twice the one seen experimentally; the discrepancy is probably to be attributed to the use of free ion charge densities vs. the ones of bound ions (present case).

In conclusion we have presented calculations for He, Ar, Kr, Xe and CH_4 interacting with the (001) surface of MgO. They agree with experimental data coming both from He beam scattering and neutron scattering experiments. The adatom-surface and adatom-adatom potentials are just the first building blocks to calculate structural or thermodynamic properties of adlayers on surfaces. To see how this can be done, the reader is referred to Ref. [41] for the case of CH_4 on graphite. We are following a similar path for CH_4 on MgO and for calculating multilayer properties.

4. Acknowledgments

We would like to thank Professor Milton Cole for illuminating discussions. The support of the American Chemical Society (Petroleum Research Fund), Research Corporation (Cottrell Research Award) and Alfred P. Sloan Foundation is gratefully acknowledged.

5. References

1. J.P. Coulomb, T.S. Sullivan, and O.E. Vilches, Phys. Rev., B30, 4753 (1984).
2. M. Bienfait, J.P. Coulomb, and J.P. Palmari, Surf. Sci., 182, 557 (1987).
3. G.P. Boato, P. Cantini, C. Guidi, R. Tatarek, and G.P. Felcher, Phys. Rev. B, 20, 3957 (1979).
4. K.H. Rieder, Surf. Sci., 118, 57 (1982).
5. G. Brusdeylins, R. B. Doak, J.K. Skofronick, and J.P. Toennies, Surf. Sci., 128, 191 (1983).

6. P. Cantini, E. Cevasco, Surf. Sci., 148, 37 (1984).
7. Here we concentrate on the behaviour of adsorbed gases at low coverage, i.e. less than a monolayer. In a subsequent paper we will consider the multilayer regime.
8. See for examples papers presented at the Symposium on Statistical Mechanics of Adsorption, Trieste, Italy, 1982 (Surf. Sci., 125 1 (1983)).
9. J.P. Coulomb, K. Madih, B. Coset, and H.J. Lauter, Phys. Rev. Lett., 54 (1985) and references cited therein.
10. V. Celli, D. Eichenauer, A. Kaufhold and J.P. Toennies, J. Chem. Phys., 83, 2504 (1985).
11. A. Frigo, F. Toigo, M.W. Cole, and F.O. Goodman, Phys. Rev. B33, 4184 (1986).
12. J.M. Hutson and P.W. Fowler, Surf. Sci., 173, 337 (1986).
13. M. Miglio, F. Quasso, and G. Benedek, J. Chem. Phys., 83, 417 (1985).
14. W.E. Carlos and M.W. Cole, Surf. Sci., 91, 339 (1980).
15. W.A. Steele, Surf. Sci., 36, 317 (1973).
16. H. Jonsson and J.H. Weare, Faraday, Dis. Chem. Soc., 80, 1 (1985).
17. G. Vidali and M.W. Cole, Phys. Rev. B29, 6736 (1984).
18. K.T. Tang and J.P. Toennies, J. Chem. Phys., 80, 3726 (1984).
19. H.U. Finzel, H. Frank, H. Hoinkes, M. Luschka, H. Nahr, H. Wilsch, and U. Wanka, Surf. Sci., 49, 577 (1975).
20. J.M. Rogowska, J. Chem. Phys. 68, 3910 (1978).
21. K.S. Petzer, Advan. Chem. Phys., 2, 59 (1959).
22. S. Rauber, J. R. Klein, M. W. Cole, and L. W. Bruch, Surf. Sci., 123, 173 (1982).
23. J.Q. Broughton, and P.S. Bagus, Phys. Rev. B, Preprint.
24. P.W. Fowler, and P.S. Madden, J. Phys. Chem., 89, 2581 (1985).
25. M. Tosi, Solid State Phys., 16, 1 (1964).
26. J.K. Norskov, Phys. Rev., B26, 2785 (1982).
27. N.D. Lang and J.K. Norskov, Phys. Rev., B27, 4612 (1983).
28. M. Karimi, G. Vidali, Phys. Rev. B36, 7576 (1987).
29. M. Manninen, J.K. Norskov, M.J. Puska, and C. Umrigar, Phys. Rev., B29, 2314 (1984).
30. G.V. Valat, J.P. Vidal, and K.K. Suonio, Acta Crys., A34, 594 (1978).
31. I.S. Gradshteyn and I.M. Ryzhik, Table of Integrals, Series and Products, (Academic Press, 1965) P. 953.
32. M.W. Cole and F. Toigo, Phys. Rev B31, 727 (1985).
33. M. Karimi, G. Vidali, Phys. Rev., B34, 2794 (1986).
34. G. Vidali, M.W. Cole, Phys. Rev., B22, 4661 (1980).
35. J.A. Barker, R.A. Fisher, and R.O. Watts, Mol. Phys., 21, 657 (1971); J.A. Barker, R.O. Watts, J.K. Lee, T.P. Schafer, and Y.T. Lee, J. Chem. Phys., 61, 3081 (1974); J.A. Barker and D. Henderson, Rev. Mod. Phys., 48, 587 (1976), Table I,P.595.
36. K. Nath, Z.W. Gortel, and H.J. Kreuzer, Surf. Sci., 155, 596 (1985).
37. D. McLachlan, Mol. Phys., 7, 381 (1964).
38. R.M. Aten, D. Blanchard, L.R. Allen, E.R. Conrad, M. Nelson, and T. Engel, Surf. Sci. 183, 515 (1987).
39. G. Vidali, C. Hutchings, and M. Karimi, in preparation.
40. R. Gomer in "Surface Mobilities on Solid Materials", Ed. by Vu Thien Binh, NATO ASI Series (Plenum, New York 1981), P.7.
41. J.M Phillips and M.D. Hammerbacher, Phys. Rev. B, 29, 5859 (1984); J.M. Phillips, ibid 29, 5865 (1984).

Comparison of Equilibrium and Non-Equilibrium Diffusion Measurements for W(110)p(2×1)-O

*M.C. Tringides**, *P.K. Wu*, and *M.G. Lagally*

Materials Science Center,
University of Wisconsin-Madison,
Madison, WI 53706, USA

The thermodynamic behavior of an overlayer on a surface is a result of its microscopic interatomic interactions. Measurements of microscopic interactions can be made at equilibrium or as the system evolves in time to attain it. The barriers to diffusion can be measured in the two cases and the atomic interactions determined. For the nonequilibrium experiment it is necessary to define a non-equilibrium diffusion coefficient that depends on the mean local environment that a diffusing atom sees. Experiments will be presented on the W(110)p(2x1)-O system. Although the macroscopic activation energies to diffusion are different in the two cases, the same set of atomic interactions is extracted. The difference is related to the different local environment that an atom "sees" in an equilibrium and a non-equilibrium experiment.

1. Introduction

Considerable effort has recently been made to identify the physics that governs systems at strongly non-equilibrium conditions [1]. A system prepared in an initial state is suddenly quenched to a state far from equilibrium. There is a well-defined final state, however, and the system will evolve in time to attain it. By studying such processes the principles governing non-equilibrium phenomena can be understood. If information about the energetics of a system is needed, it is more common to study a system after it has achieved equilibrium. It is a state of minimum free energy, with the only deviations from it being small thermal fluctuations. The laws of thermodynamics correlate the observable macroscopic parameters to the atomic microscopic interactions. However, as will be discussed later, information about the energetics can be also obtained from a non-equilibrium quench experiment.

In statistical mechanics one usually specifies a set of a few, local interaction parameters and some underlying lattice geometry and the full thermodynamic behavior of the system in principle follows. In actuality, of course, one faces the opposite problem: macroscopic parameters are first measured and after appropriate modeling the microscopic interactions are determined. For example, one can map out the (T,Θ) phase boundary from a diffraction experiment, where T is the temperature and Θ the coverage. From the symmetry of the low-temperature ordered phase the interaction range and their sign are obtained. The magnitude is deduced from a more quantitative modeling of the phase boundary. Other experiments that provide information about the interactions in the system are the study of activated processes, such as the dependence of diffusion rate on coverage. Such experiments [2,3] have been performed at equilibrium for W(110)p(2x1)-O and the microscopic interactions determined [4,5]. In this paper we will show that non-equilibrium quench experiments [6,7] done on the same system suggest the same set of atomic interactions. However, the activation energies measured are

*Present address: Department of Physics and Ames Laboratory-USDOE, Iowa State University, Ames, Iowa 50011

different because, although the same set of atomic interactions are applicable, different atomic configurations contribute to the statistical average in equilibrium vs non-equilibrium experiments.

The outline of the paper is as follows: in Section 2 we describe the experiments for measuring the equilibrium phase diagram with LEED and equilibrium diffusion coefficients with the current fluctuation method for the W(110)p(2x1)-O system. In Section 3 we present the results of a non-equilibrium experiment on the same system. In Section 4 we explain how the measured activation energies lead to the same microscopic interactions between adsorbate-adsorbate and substrate-adsorbate atoms in the two experiments. We also comment how the above considerations are relevant to MBE growth experiments [8] where diffusion is measured from the temperature dependence of RHEED oscillations in the specular intensity. We end with a short conclusion.

2. Equilibrium Measurements

2.1 Phase Diagram Determination

The O/W(110) phase diagram in the temperature vs coverage plane has been mapped [2] out with the use of LEED. After the exposure to a known amount of oxygen at $T \simeq 160$ °K and annealing at a higher temperature (~ 800°K) well ordered structures are formed at room temperature. This is evidenced by the appearance of superstructure spots on the diffraction pattern. Below $\Theta = 0.5$ the ordering corresponds to the formation of the p(2x1) structure. Two equivalent domains are possible, rotated from each other by the angle of the unit cell. Above $\Theta = 0.5$ the diffraction pattern suggests the presence of p(2x1) and p(2x2) ordered phases until eventually at one monolayer a p(1x1) pattern is recovered. We will limit our discussion to coverages less than 0.5 because at higher coverages the presence of two ordered phases complicates the analysis. If the overlayer is raised to a higher temperature (for sufficiently long time so it equilibrates) the diffraction spots become weaker. Part of the reduction is due to the Debye-Waller factor: the thermal vibrations introduce deviations in the positions of the scattering atoms and incoherence in the diffracted intensity. However, the influence of the Debye-Waller factor can be eliminated because it decays exponentially with temperature. A semilog plot of peak intensity vs temperature should be a straight line if only Debye-Waller effects are present. Experimental measurements, however, show stronger deviations from such exponential decay, which suggests the presence of another disordering mechanism. A typical experimental curve of the decay of the intensity is shown in Fig. 1 for the (0,1/2) beam after the Debye-Waller contribution is eliminated. It is clear that the decrease of intensity signals the loss of long range order in the p(2x1) overlayer. In the same curve the corresponding full width at half maximum of the measured angular profiles (corrected for the instrument width) is shown as a function of temperature. It increases with temperature, also suggesting that the p(2x1) order decreases. A characteristic temperature can be obtained from the above curves at their inflection point, signaling the order-disorder transition. Such intensity-vs-temperature curves can be obtained at different coverage and inflection points specified so the phase diagram boundary separating the ordered from the disordered phase, can be identified on the (T,Θ) plane.

In Fig. 2 the experimentally determined phase diagram is shown schematically. It is rather likely, however, that the low-temperature region corresponds to a coexistence between the p(2x1) and a disordered lattice gas phase bounded by a line of first-order transitions. In the higher-temperature (480° < T < 800°K) region a single p(2x1) ordered phase is stable above a critical coverage. The phase region is bounded by a line of second-order transitions. This identification is based on the model microscopic

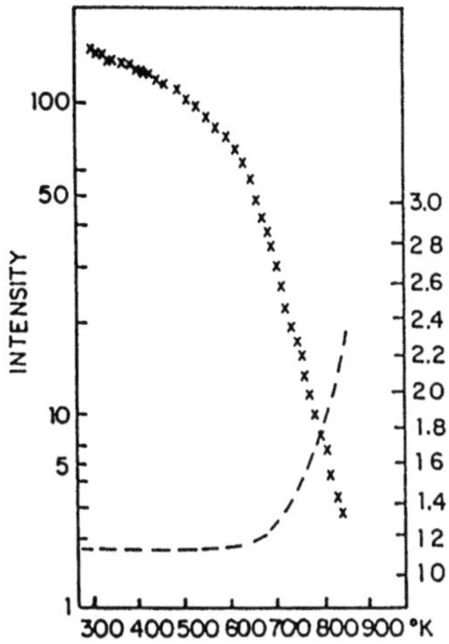

Fig. 1. Temperature dependence of the (1/2,0) beam intensity (xxxx) and profile angular width at half maximum (----) for well equilibrated p(2x1) - O over-layer on W(110) at $\Theta = 0.5$. The transition temperature is determined by the inflection point in the intensity curve (from Ref. 2)

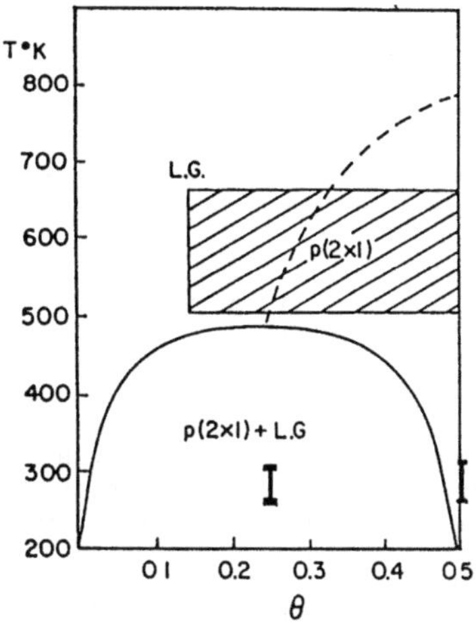

Fig. 2. Experimentally determined phase diagram for W(110)p(2x1) - O. The solid line is a line of first-order and the dashed one of second-order transitions. The equilibrium diffusion experiments were done within the hatched region. The quench temperatures for the non-equilibrium experiments are denoted by the heavy lines

interactions that are needed to produce the observed p(2x1) phase as the ground state of the system. It can be shown from calculation of the energy at zero temperature that to have p(2x1) as the ground state, essentially two microscopic interactions are needed, a nearest neighbor attraction, ϕ_1, and a next-nearest repulsion, ϕ_2, such that $|\phi_2| \geq \frac{|\phi_1|}{2}$. Theoretical analysis [9] of such models has shown that the phase diagram consists of a low-temperature coexistence region with a second-order line at higher temperatures. The above analysis is correct for $\Theta < 0.5$; if the whole phase diagram up to $\Theta = 1$ is modeled, triplet interactions are needed [10]. There is also experimental evidence that the above identification of the phase boundaries and the sign of the interactions is justified: at the lower temperatures and coverages it is quite unlikely to have a p(2x1) single-phase region, because the presence of so many vacancies requires the formation of regions of p(2x2) phase, which is never observed. More information can be obtained about the microscopic interactions ϕ_1, ϕ_2 if not only the ground state symmetry is considered but the actual values of the measured transition temperatures are fitted. Monte Carlo simulations to fit the decay of the (1/2,0) intensity with temperature suggest $|\phi_1| \simeq |\phi_2| \simeq 0.09$ eV, and also include a second-nearest-neighbor interaction ϕ_3 along the closed-packed direction with $\phi_3 \simeq \phi_1/3$. The interaction ϕ_3 is not needed to establish p(2x1) as the ground state and it is not clear if it is necessary for the intensity fit, because any comparison between the simulation and the experiment requires analysis of finite-size effects.

2.2 Equilibrium diffusion measurements

Measurements of time-dependent processes on surfaces, especially at equilibrium, are limited. A microscopic field emission method that utilizes the high sensitivity of the field emission current to the presence of adsorbates has been applied to diffusion studies. A full description of the method and applications to several experimental systems can be found elsewhere [11,13]. Here we will give a brief summary and the results of measurements performed on the W(110) p(2x1)-O system.

A field emitter of radius approximately 2000 Å with <110> orientation exposes normal to its axis a (110) plane of area A_t with roughly 1000 Å dia. A submonolayer coverage of oxygen is dosed onto the emitter and the system is well equilibrated. The amount of oxygen can be monitored by measuring the work function. When the work function remains constant during the experiment, it implies that the oxygen coverage does not change. The measurement is on a well equilibrated layer under constant-coverage conditions. The field emission current originating from a central probe region of area A on the plane is monitored as a function of time. A is chosen so it is much less than the total area of the plane A_t ($A \leq A_t/5$). As the adsorbate atoms move in and out of the probe area, they cause changes in the "effective" work function of the area and therefore fluctuations in the field emission current. As the temperature is raised and the adsorbate diffuses faster the fluctuations will also become faster. If n(t) denotes the number of atoms at time t, $\bar{n}$ the average number of atoms, and i(t), $\bar{i}$ the corresponding values of the field emission current in the probe area, it can be shown that

$$\delta n(t) = (n(t)-\bar{n}) \propto \delta i(t) = i(t) - \bar{i} \ . \tag{1}$$

Therefore, the autocorrelation functions formed for the quantities in Eq. (1) are proportional to each other:

$$f_n(t) = \langle \delta n(t)\ \delta n(0) \rangle \propto \langle \delta i(0)\ \delta_i(t) \rangle = f_i(t) \ . \tag{2}$$

The function $f_n(t)$ qualitatively behaves like a decaying exponential because an initial fluctuation will eventually be uncorrelated with fluctuations occurring at very long times. The characteristic decay time t_o is inversely proportional to the diffusion coefficient. Experimentally one measures the decay of the field emission current fluctuation $f_i(t)$ but, because it is proportional to $f_n(t)$, the characteristic time t_o is obtained. This time corresponds to decay of the adsorbate fluctuations across an area A. Hence, the diffusion coefficient is obtained from

$$D = \frac{A}{4t_o} \ . \tag{3}$$

The probe area A is found by utilizing the magnification of the microscope and the field emitter radius. The above analysis has been shown [12] to be valid also for the case where adsorbate-adsorbate interactions are present.

The O/W(110) system has been studied in the coverage range $0.15 < \Theta < 0.62$, both with a circular probe area so an average diffusion coefficient is obtained [3a] and with a rectangular one, so the anisotropy in diffusion is measured [3b]. A typical Arrhenius plot for $\Theta = 0.62$ is shown in Fig. 3. We limit our discussion here to the variation of activation energy with coverage. Similar activation energies are obtained for both probe geometries. The value is 0.6 eV at $\Theta - = 0.15$ and then increases up to 1 eV at $\Theta = 0.62$. Figure 4 shows the change with coverage. A rapid increase occurs within a very narrow range in coverage around $\Theta = 0.3$. Similar behavior with coverage is seen for the prefactor. It has a rather low value of 10^{-7} cm^2/sec at $\Theta = 0.15$ and increases to a more normal value of 10^{-4} cm^2/sec at $\Theta = 0.62$. A dependence of the activation energy on coverage implies the presence of adsorbate-adsorbate interactions. The diffusion

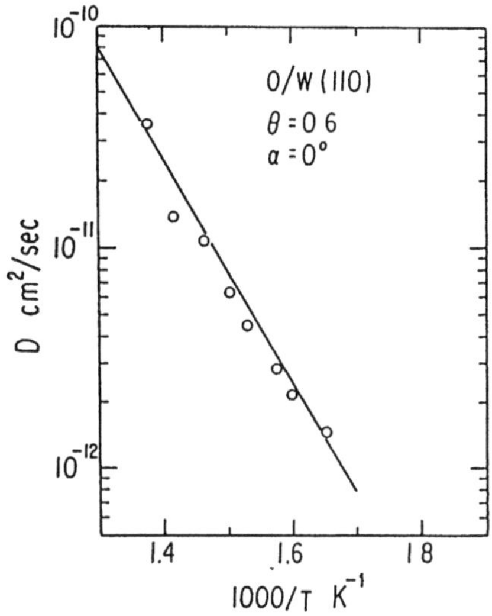

Fig. 3. Arrhenius plot for diffusion coefficient vs 1/T for W(110)p(2x1) - O at equilibrium. The coverage is 0.62. The uncertainty in the data is approximately 15% (from Ref. 3b)

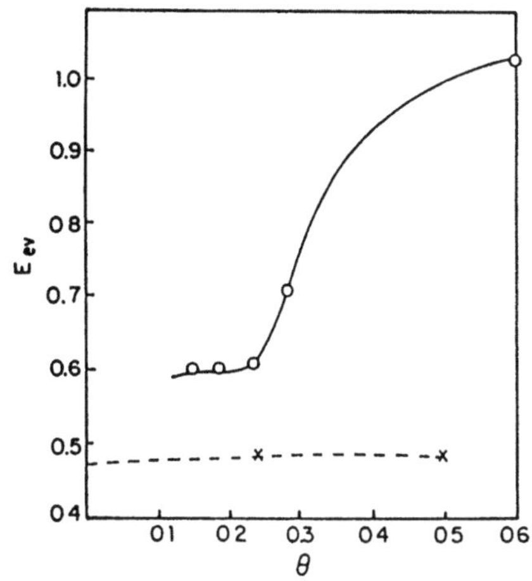

Fig. 4. Variation of diffusional activation energies with coverage. The points (o) denote the results of the equilibrium experiment and (x) those at non-equilibrium. As discussed in the text, the equilibrium values are higher than the non-equilibrium ones because they correspond to diffusion in the p(2x1) phase

barrier is the sum of the substrate-adsorbate potential plus changes due the adsorbate-adsorbate interactions. A rising activation energy with increasing coverage suggests that a least some of the interactions must be attractive. The exact value of the activation barrier, for a given set of microscopic pairwise interactions, will depend on the local configuration.

The range in temperature over which these diffusion measurements were made is shown in the cross-hatched region in Fig. 1. Lower temperatures than these are excluded because at low temperatures the non-diffusive signal from the substrate dominates the fluctuations. Higher temperatures were inaccessible because the time scale then is faster than the response of the electronics. It is clear that the low-coverage measurements were performed in the single-phase lattice gas region and the higher-coverage ones in the p(2x1) ordered region. As argued in Ref. 5, this explains the sudden rise at $\Theta = 0.3$ of both the activation energy and the prefactor. A fit to the results was obtained with interactions used in Ref. 4 [$\phi_1 \simeq \phi_2 \simeq 0.1$ eV] (a better fit was obtained with the use of an additional triplet repulsive interaction $\phi_t = -0.052$ eV but this term is important only for $\Theta > 0.5$). Consequently, the equilibrium measurements of both the phase diagram and activation energies of diffusion can be explained with the same set of microscopic interactions.

3. Non-Equilibrium measurements

As stated in the introduction it is of interest to study overlayers not only at equilibrium but also in non-equilibrium situations. A non-equilibrium situation is easily realized after a quench of one of the thermodynamic variables (temperature, chemical potential). Instantaneously the variable acquires a new value, in a different region of the phase diagram in which the system has a new equilibrium state. The system will evolve in time to attain this state. Experimentally such phenomena are studied with the use of diffraction because the change of symmetry can be easily monitored through changes in the diffracted beam. Important considerations in such experiments were presented elsewhere [13,14]. Such experiments [6] have been performed recently.

A known amount of oxygen (at this temperature) is deposited at low temperature, T = 200°K. Because the mobility is zero at this temperature, the configuration will correspond to an infinite-temperature one. Then the surface is annealed at a somewhat higher temperature, e.g., 260-300 K. The oxygen atoms have enough mobility at these temperatures to diffuse and form ordered islands. The experiment was performed at $\Theta = 0.25$ and $\Theta = 0.5$; the range in temperature for $\Theta = 0.25$ lies within the coexistence region and for $\Theta = 0.5$ within the single phase region.

Results on the growth law and the domain size distribution have been presented elsewhere [6,7]. Briefly, the domain size distribution is time invariant, hence growth is self-similar. The growth of the average domain size, L, obeys a power law,

$$L \approx A(T)t^{x} , \qquad (4)$$

with the exponent $x \approx .28$ at both coverages, in good agreement with theoretical expectations [15,16]. Here we will focus on the temperature dependent growth rate A(T). An Arrhenius plot for $\Theta = 0.5$ is shown in Fig. 5. The activation energy is approximately 0.16 eV and a similar value is found at $\Theta = 0.25$. The value is related to the diffusional barrier. Only two model calculations of A(T) exist. First, a system with two ground states and no conservation law for the number of particles is described by the Lifshitz-Cahn-Allen law,

$$L = \left(\frac{Dnv}{kT} t \right)^{1/2} , \qquad (5)$$

where v is interaction energy, n the order parameter, k Boltzmann's constant, and T the temperature. Second, phase separation in the coarsening stage is described by the Lifshitz-Slyozov law,

$$L = \left(\frac{8\gamma DCV_m t}{9RT} \right)^{1/3} , \qquad (6)$$

where γ is the surface energy, C the concentration, V_m the molar volume, and R the gas constant. In both of these results the diffusion coefficient appears with the same exponent as the time. This can be generalized for any growth law if we assume that the only time dependence in the right hand side of Eq. (4) is contained in D and t, which is true for Eqs. (5) and (6). Then, since D has units of inverse time it must have the same exponent as the time to cancel the overall time dependence in the rhs of Eq. (4). It follows

$$A(T) \propto D^{x} \qquad (7)$$

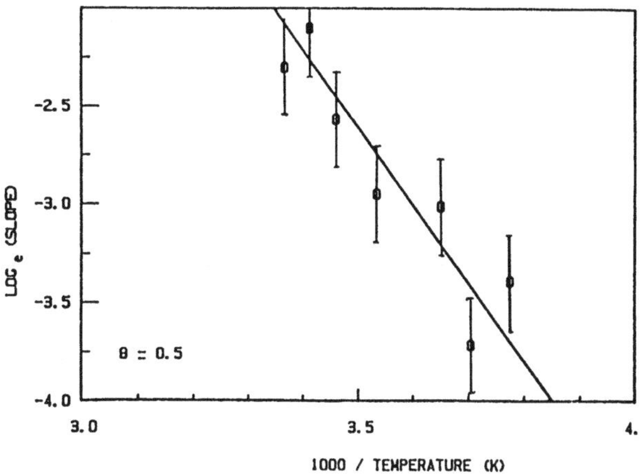

Fig. 5. Arrhenius plot for the growth rate A(T) vs 1/T at $\Theta = 0.5$ for the non-equilibrium experiment. The activation energy determined is n times the activation energy of diffusion where n = 0.28 is the growth exponent

and the diffusional barrier is 1/x times the activation energy measured for A(T). Therefore, the activation energy for diffusion is approximately 0.5 eV for both $\Theta = 0.25$ and $\Theta = 0.5$. There is no reason for the activation energies to be the same at these two different coverages because more atoms contribute to the local interactions at $\Theta = 0.5$ than do at $\Theta = 0.25$ unless some special condition is satisfied. This will be discussed further in the next section. If we also compare these non-equilibrium values to the values of the activation energy measured at equilibrium we see that they are consistently lower, especially at $\Theta = 0.5$.

4. Discussion

If there are no adsorbate-adsorbate interactions then it is clear that the barrier to diffusion will be the substrate-adatom potential with no coverage dependence. The presence of atomic interactions within the overlayer will make the activation energy coverage dependent, as observed experimentally. The effective barrier to diffusion at any coverage will be determined both by the microscopic interactions and the most probable atomic configurations. These are drastically different in the equilibrium and the non-equilibrium experiments. In the first case the whole surface is predominantly covered with p(2x1) phase, and so an oxygen atom on the average will "see" only the attractive interactions due to its nearest neighbors. This will tend to raise the effective single-particle diffusional activation energy. In the second case the typical configuration is expected to be random and so on the average an atom will "see" the same number of nearest and next-nearest neighbors. Therefore, in this configuration, both attractive and repulsive interactions are present. We can crudely write

$$E(\Theta) = E(0) + Z\Theta(\phi_1 - \phi_2) \qquad (8)$$

from a simple bond counting argument, where Z is the coordination number of the lattice, and $E(\Theta)$ the activation energy at finite coverage. Using the condition $|\phi_1| \simeq |\phi_2|$ obtained in fits to the phase diagram, we find for the non-equilibrium experiment that $E(\Theta) \simeq E(0)$, independent of coverage.

Hence, the measured activation energy at both $\Theta = 0.25$ and $\Theta = 0.5$ is essentially the depth of the substrate-adatom potential well. The slightly higher value of $E = 0.6$ eV at $\Theta = 0.15$ in the equilibrium measurement is consistent with this picture. Although at this low coverage the equilibrium phase is disordered, fluctuations can create very small regions of p(2x1), 3-4 atoms wide, formed locally. The impact of these fluctuations-induced regions is to raise the activation energy. When the p(2x1) phase becomes stable, the activation energy rises rapidly.

The above analysis is, of course, only approximately correct. The complete randomness of the configurations is true only at zero time. A more accurate analysis will require proper weighting of each configuration with the correct probability factor, necessitating the use of Monte Carlo methods or an approximation at a more sophisticated level. The equilibrium case has been treated[5] by Monte Carlo methods and it was shown, as stated previously, that indeed the formation of the p(2x1) phase leads to the rise of the activation energy seen experimentally.

Non-equilibrium growth kinetics measurements have also been made for the W(211)p(2x1)-O system at $\Theta = 0.5$ [17]. An activation energy for the growth rate A(T) of $E = 0.14$ eV was found. The growth exponent of $x = 1/2$ implies a diffusional activation energy of $E = 0.28$ eV. Equilibrium experiments [18] with the fluctuation method on a (211)-oriented field emitter measured $E = 0.61$ eV at $\Theta = 0.5$ (two distinct temperature regions are seen in this experiment, with different activation energies; only the high-temperature one is identified as diffusion). Again the non-equilibrium value is less than the equilibrium one, reflecting the randomness of the atomic arrangements. This result should be general because in an equilibrium experiment a state of minimum free energy is established and therefore a much higher mobility barrier exists than for true single-particle diffusion.

We would like briefly to address a different experiment [8] that has been used to measure diffusion and extract single-particle diffusion activation energies. This is related to MBE growth and the observations of oscillations in the diffracted-beam intensity during growth. Intensity oscillations are caused by layer-by-layer growth. As material is deposited on a flat surface the intensity at appropriate diffraction conditions drops because of cluster or island formation by the deposited atoms, causing destructive interference between the "up" level and the "down" level of atoms. It reaches a minimum at half coverage and then increases again as the layer becomes complete. Similar phenomena have been observed on stepped surfaces. As the temperature is raised the amplitude of the oscillations decreases, although the period remains constant. This phenomenon is explained by the increased mobility of the atoms at higher temperature, and their eventual capture by the step edge. This mechanism will lead to no oscillations at the condition that the arrival rate R matches the capture rate by the step. For a [Cgiven incident flux a temperature T is determined at which such condition is satisfied and a plot of R^{-1} vs T^{-1} will give the activation energy. This has been identified as the single-particle diffusion activation energy but such a picture is too simplistic. Even if only single atoms were present on the surface, their motion is not a true random walk. The presence of the capturing step edge will bias the motion in one direction, and one measured diffusion under a gradient and the measured activation energy will have a contribution due to this gradient.

Moreover, it is not likely that only isolated atoms exist on the surface. The experiment [13,14] is basically a non-equilibrium one, with the chemical potential now being the quench variable. The surface is in a dynamic situation in which cluster formation competes with diffusion towards the step. The elimination of the intensity oscillations suggests only that a steady state has been reached where on the average the distribution of atoms on the terrace is invariant because as the new atoms arrive from the source they are balanced by the atoms diffusing towards the step. In addition, clusters could have their own mobility. The presence of islands can be determined from a careful analysis of the profile shape. The degree to which these remaining clusters affect the diffusion coefficient can only be addressed by modeling the dynamic system with appropriate interactions.

5. Conclusion

Three different experimental measurements on W(110) p(2x1) - O, done over a span of 12 years, performed on different experimental apparatus, and probing different aspects of the system, were presented. The first two experiments probe the system at equilibrium and the other one far from equilibrium. The measured macroscopic phenomenological activation energies for diffusion were different. This, however, results from the nature of the experiment itself, where different microscopic configurations are sampled. All the experiments suggest the same set of microscopic interactions and a separation of the substrate-adsorbate and adsorbate-adsorbate part is possible. It is clear that one has to discuss separately a diffusion coefficient at equilibrium and a non-equilibrium one. The first one is well defined and should be the same irrespective of the method used to probe the system. For interpreting the second one, however, care is needed to characterize the non-equilibrium nature of the system. Useful information about the basic symmetries of the system and the microscopic interactions can be extracted by comparing such measurements.

Acknowledgment

This research was supported by NSF, Solid State Chemistry Program, under Grant No. DMR 86-15089. Support for MCT during the preparation of the manuscript was provided by Iowa State University and Ames Laboratory. Ames Laboratory is operated for the U.S. Department of Energy by Iowa State University under contract no. W-7405-Eng-82.

6. References

1. J. D. Gunton, M. San Miguel, and P. S. Sahni: in Phase Transitions and Critical Phenomena, Vol. 8, p. 267 (1983) ed. by C. Domb and J. Lebowitz (New York, Academic Press)
2. G. C. Wang, T. M. Lu, and M. G. Lagally: J. Chem. Phys. 69, 479 (1978)
3. a) J. R. Chen and R. Gomer: Surf. Sci. 79, 413 (1979); b) M. Tringides and R. Gomer: Surf. Sci. 155, 254 (1985)
4. E. D. Williams, S. C. Cunningham, and W. H. Weinberg: J. Chem. Phys. 68, 5 (1978); W. Y. Ching, D. L. Huber, M. Fishkis, and M. G. Lagally: J. Vacuum Sci. Techn. 15, 613 (1978)
5. M. Tringides and R. Gomer: Surf. Sci. 45, 121 (1984)
6. M. Tringides, P. K. Wu, and M. G. Lagally: Phys. Rev. Lett. 59, 315 (1987)
7. P. K. Wu, M. Tringides, and M. G. Lagally: in preparation
8. J. H. Teague, P. J. Dobson, R. A. Joyce, and J. Zhang: Appl. Phys. Lett. 47, 100 (1985); P. R. Pukite, S. Batra, and P. I. Cohen: Proceedings of NATO Advanced Workshop on "RHEED and Reflection Electron Imaging of Surfaces," Plenum, 1988 (in press)
9. K. Binder and D. P. Landau: Phys. Rev. B 21, 1941 (1980)
10. W. Y. Ching, D. L. Huber, M. G. Lagally, and G. C. Wang: Surf. Sci. 77, 550 (1978)
11. R. Gomer: Surf. Sci. 38, 373 (1973)
12. G. Mazenko, J. R. Banavar, and R. Gomer: Surf. Sci. 107, 459 (1981)

13. M. C. Tringides and M. G. Lagally: "Studies of Growth Kinetics on Surfaces with Diffraction," in Proceedings of NATO Advanced Workshop on "RHEED and Reflection Electron Imaging of Surfaces," Plenum, 1988 (in press)
14. M. G. Lagally, D. E. Savage, and M. C. Tringides: "Diffraction from Disordered Surfaces: an Overview," in Proceedings of NATO Advanced Workshop on "RHEED and Reflection Electron Imaging of Surfaces, Plenum, 1988 (in press)
15. A. Sadiq and K. Binder: J. Stat. Phys. 35, 617 (1984)
16. I. M. Lifshitz and V. V. Slyozov: J. Chem. Phys. Solids 15, 35 (1961)
17. G. C. Wang and T. M. Lu: private communication
18. M. C. Tringides and R. Gomer: J. Chem. Phys. 84, 7 (1986)

Hopping and Diffusion

H.J. Kreuzer

Department of Physics, Dalhousie University,
Halifax, N.S. B3H 3J5, Canada

Diffusion is the macroscopic mass transport that results from gradients in the chemical potentials of the various components of a fluid or solid mixture. It is controlled by Fick's law which reads for a one component system

$$\frac{\partial n(\mathbf{r},t)}{\partial t} - \frac{\partial}{\partial \mathbf{r}} \left(D \frac{\partial n(\mathbf{r},t)}{\partial \mathbf{r}}\right) = 0 \tag{1}$$

Here $n(\mathbf{r},t)$ is the number density of diffusing particles and $D=D(n)$ is the diffusion coefficient which can, in general, be density dependent. Generalizations of (1) to multi-component systems with the inclusion of thermal effects and other external forces such as electric fields have been formulated within the framework of nonequilibrium thermodynamics and can be found in standard textbooks on the matter. Diffusion in solids, and more generally atomic transport in solids, has been reviewed recently by ALLNATT and LIDIARD /1/.

It is the role of microscopic models to calculate the diffusion coefficients of a given system from first principles, i.e. from the microscopic dynamics of the constituent atoms, molecules or ions of the system. This has been done successfully for dilute gases starting from the Boltzmann equation. A general approach has been formulated in linear response theory. We want to discuss here a simple microscopic model for diffusion on surfaces that has been successfully studied for diffusion, in particular of hydrogen, in crystalline alloys /1/. Its application to surface diffusion has been discussed in some detail by BOWKER and KING /2/ who also presented extensive computer simulations.

The model considers the motion of particles within a rigid matrix. On a surface the latter is given by the set of localized adsorption sites which we label by an index "i". We want to formulate the model for two kinds of adsorbed particles, called "n" and " p", e.g. two isotopes of the same species such as H_2 and D_2. We introduce site occupation numbers such that $n_i=1$ if an n-particle occupies site i and $n_i=0$ if site i is not occupied by an n-particle; likewise $p_i= 1$ or 0 if a p-particle occupies site i or not. Obviously we have $n_i+p_i=0,1$. Next we assume that a particle of either kind can jump to a neighbouring site provided the latter is empty. This blocking is the result of strong hard core repulsion; any further interaction between particles is neglected. The jumps between sites are envisaged to be instantaneous, i.e. of a duration much shorter than the residence time in a site. Moreover, it is assumed that hopping is a Markov process, i.e. that the jumps at any given time are independent of each other and of the past history of the system. This defines the hopping model with site exclusion.

Proceeding to a mathematical formulation of the hopping model, we introduce a function

$$P(n_1,p_1;n_2,p_2;\ldots;n_i,p_i;\ldots;t) = P(\mathbf{n},\mathbf{p};t) \qquad (2)$$

that gives the probability that at a given time t any one of the sites i are occupied by an n-particle ($n_i=1$) or by a p-particle ($p_i=1$) or not at all ($n_i=p_i=0$). Thus $P(\mathbf{n},\mathbf{p};t)$ is a function of 2N+1 variables and operates in a 2^{2N} dimensional phase space. Its time evolution is controlled by a master equation, valid for any Markov process,

$$\frac{\partial P(\mathbf{n},\mathbf{p};t)}{\partial t} = \sum_{\mathbf{n}',\mathbf{p}'=0,1} [W(\mathbf{n},\mathbf{p};\mathbf{n}',\mathbf{p}')P(\mathbf{n}',\mathbf{p}';t) - W(\mathbf{n}',\mathbf{p}';\mathbf{n},\mathbf{p})P(\mathbf{n},\mathbf{p};t)] \qquad (3)$$

Here $W(\mathbf{n},\mathbf{p};\mathbf{n}',\mathbf{p}')$ is the probability per unit time that the system, initially in a microscopic state specified by the occupation numbers $(\mathbf{n}',\mathbf{p}')$ will make a transition to a state specified by a set $(\mathbf{n},\mathbf{p})$. A thorough discussion of this equation is given in /3/ and /4/; the latter monograph also gives details on how to calculate W from a microscopic hamiltonian for the case of phonon assisted adsorption-desorption kinetics. For the hopping model specified above we have

$$W(\mathbf{n}',\mathbf{p}';\mathbf{n},\mathbf{p}) = \sum_{i,j(n.n.)} [J\, p_i(1-n_j-p_j)\, \delta_{p_i',1-p_i}\, \delta_{p_j',1-p_j}\, \delta_{n_i',n_i}\, \delta_{n_j',n_j}$$

$$+ J_0\, n_i(1-n_j-p_j)\, \delta_{n_i',1-n_i}\, \delta_{n_j',1-n_j}\, \delta_{p_i',p_i}\, \delta_{p_j',p_j}] \prod_{k\neq i,j} \delta_{p_k',p_k}\, \delta_{n_k',n_k} \qquad (4)$$

The first term is nonzero if a p-particle occupies initially site i ($p_i=1$) and hops, with a rate J per unit time, to an empty ($n_j+p_j=0$) neighbouring site j so that in the final state $p_i'=0$ and $p_j'=1$. The second term gives the analogous rate for a n-particle with a rate constant J_0. The other δ-functions account for the fact that all other sites remain in their original occupation. The sum in (4) runs over all pairs of nearest neighbour sites. Further interactions, in addition to the blocking repulsion, would add additional terms in (4); they have been worked out in Ising models and for adsorption-desorption kinetics, see e.g. /5,6/. We stress that the hopping probability of a particle is independent of its environment, i.e. of the occupation of all other sites. This, in particular, implies that we do not account for the possibility that two particles will attempt to hop to the same final site; this is a consequence of the fact that hops are assumed to be instantaneous.

Because the transition probabilities (4) satisfy detailed balance, the solution of (3), starting from some initial condition, will evolve towards equilibrium. Solving (3) by straightforward matrix diagonalisation is not feasible due to the large size of the matrices involved. Quite accurate Monte Carlo simulations have been advanced /7-9/. To get some insight into the physics of the problem we have developed a scheme by which one can easily extract the long time, i.e. diffusion-like, behaviour. We start by defining mean occupation numbers

$$\langle n_i(t)\rangle = \sum_{\mathbf{n},\mathbf{p}=0,1} n_i\, P(\mathbf{n},\mathbf{p},t) \qquad (5)$$

Note that these mean values take on any value between 0 and 1, whereas the microscopic variables n_i and p_i take on values 0 and 1 only. They are still functions of time and satisfy a set of coupled differential equations, derived by averaging (3) according to (5), namely

$$\frac{d}{dt}\langle n_i(t)\rangle = J_0 \sum_a [\langle n_{i+a}\rangle(1-\langle p_i\rangle) - \langle n_i\rangle(1-\langle p_{i+a}\rangle) - \langle \nu_{(i+a)}\pi_i\rangle + \langle \nu_i \pi_{i+a}\rangle] \tag{6}$$

$$\frac{d}{dt}\langle p_i(t)\rangle = J \sum_a [\langle p_{i+a}\rangle(1-\langle n_i\rangle) - \langle p_i\rangle(1-\langle n_{i+a}\rangle) + \langle \nu_{i+a}\pi_i\rangle - \langle \nu_i \pi_{i+a}\rangle] \tag{7}$$

Here a labels the neigbhours of site i. For ease of notation we dropped the explicit time dependences on the right hand side of these equations. Also we defined fluctuations

$$\nu_i = n_i - \langle n_i\rangle \tag{8}$$
$$\pi_i = p_i - \langle p_i\rangle \tag{9}$$

The $\langle\nu\pi\rangle$ correlations between neighbouring sites in (6) and (7) satisfy rather complicated differential equations which couple them (i) to higher correlations of the type $\langle\nu\nu\pi\rangle$ and $\langle\nu\pi\pi\rangle$, and (ii) to $\langle\nu\nu\rangle$, $\langle\nu\pi\rangle$ and $\langle\pi\pi\rangle$ correlations between next nearest neighbour pairs.

A hierarchy of Green's functions, similar to (6) and (7), has been solved approximately in various truncation schemes /10,11/. One gets the impression, however, that this model should be solvable exactly. The problem is, indeed, trivial for the case of one component. Setting all $n_i=0$, we get from (7)

$$\frac{d}{dt}\langle p_i(t)\rangle = J \sum_a [\langle p_{i+a}\rangle - \langle p_i\rangle] \tag{10}$$

It is at first sight surprising to note the absence of blocking factors. They can be made to re-appear by writing (10) equivalently

$$\frac{d}{dt}\langle p_i(t)\rangle = J \sum_a [\langle p_{i+a}\rangle(1-\langle p_i\rangle) - \langle p_i\rangle(1-\langle p_{i+a}\rangle)] \tag{11}$$

The additional terms obviously cancel suggesting that the "intuitive" picture of blocking is somewhat misleading for identical particles; it only applies to tagged particles, i.e. an alloy or a tracer.

Rather than looking at the general solution of (10) we want to look at the large time behaviour where we expect diffusive mass transport to emerge. Because diffusion is a macroscopic notion the mean site occupation functions $\langle p_i\rangle$ will only change little over nearest neighbour distances. Thus we treat them as a function of a continuous coordinate $\mathbf{x}$ and expand it in a Taylor series

$$\langle p_{i+a}(t)\rangle = p(\mathbf{x}+\mathbf{a},t) = p(\mathbf{x},t) + \mathbf{a}\cdot\frac{\partial p}{\partial \mathbf{x}} + \tfrac{1}{2}\,\mathbf{aa} : \frac{\partial^2 p}{\partial \mathbf{x}\partial \mathbf{x}} + \dots \tag{12}$$

Inserted into (11) this results in the diffusion equation (1) because the term linear in a vanishes due to symmetry. The diffusion coefficient is then identified, in an isotropic lattice, as

$$D = \tfrac{1}{2}\, J\, a^2 \tag{13}$$

Thus the diffusion coefficient is independent of density for an ensemble of identical particles hopping on a lattice with site blocking the only interaction. This seems an appropriate model to describe hydrogen diffusion on W(100) /12/. We repeat again that for such a single component a density dependence emerges if additional interactions become important because the hopping probability will then reflect the local environment of the hopping particle. We also stress that diffusion for an alloy, for which site blocking is the only interaction, is density dependent even if one of the components is present at tracer concentrations /1/.

The expression (13) is of the general form obtained by Einstein treating diffusion as the long time limit of Brownian motion, namely

$$D = \frac{\langle(\Delta x)^2\rangle}{2\Delta t} \tag{14}$$

if we identify $\Delta t = J^{-1}$ and $\Delta x = a$. This formula is the basis for a microscopic measurement of the diffusion coefficient as first brought to perfection by PERRIN /13/. For the sake of clarity we remark that in a statistical theory the average <...> in (14) is taken over an equilibrium ensemble of particles. Experimentally, one measures the displacement Δx of one (or a few) particle as a function of time Δt. Under equilibrium conditions this microscopic motion will, over long times, cover the surface uniformly establishing the fact that, in the absence of external fields, a (simple) system in equilibrium is homogeneous; this is a consequence of ergodicity. Obviously in equilibrium there is no mass transport, and thus no diffusion; the microscopic hopping motion merely produces short time fluctuations. On the other hand, Einstein's formula (14), and more generally the fluctuation-dissipation theorem, asserts that for linear transport processes close to equilibrium dissipation, i.e. diffusion, is controlled by equilibrium fluctuations, i.e. hopping; for details see e.g. /14/. However, for mass transport far from equilibrium, as it may e.g. arise after a rapid quench into a different phase, nonlinear phenomena may become important for which Einstein's formula and the fluctuation-dissipation theorem are no longer appropriate, as suggested by the experiments of TRINGIDES et al. /15/.

Acknowledgment

This work was supported in part by a grant from the Natural Sciences and Engineering Council of Canada.

References

1. A.R. Allnatt, A.B. Lidiard: Rep. Prog. Phys. 50, (1987)
2. M. Bowker and D.A. King: Surf. Sci. 71, 583 (1978); Ibid. 72, 208 (1978)
3. N.G. van Kampen: Stochastic Processes in Physics and Chemistry (North-Holland Publishing Company, Amsterdam 1981)
4. H.J. Kreuzer, Z.W. Gortel: Physisorption Kinetics, Springer Ser. in Surface Sci., Vol.1 (Springer, Berlin, Heidelberg 1986)
5. K. Kawasaki: In Phase Transitions and Critical Phenomena, Vol.2, ed. by C. Domb and M.S. Green (Academic Press, New York 1972)
6. D.J.W. Geldart, H.J. Kreuzer, F.S. Rys: Surf. Sci. 176, 284 (1986)
7. G.E. Murch: Phil. Mag. A35, 116 (1983)
8. A.R. Allnatt, E.L. Allnatt: Phil. Mag. A49, 625 (1984)
9. R.A. Tahir Kheli: Phys. Rev. B27, 7229 (1983); Ibid. 28, 3049 (1983)
10. R.A. Tahir Kheli, R. J. Elliott: Phys. Rev. B27, 844 (1983)
11. P.C.W. Holdsworth, R.J. Elliott: Phil. Mag. A54, 601 (1986)
12. S.M. George: This volume
13. J. Perrin: Les Atomes (Presses Universitaires de France, Paris 1948)
14. H.J. Kreuzer: Nonequilibrium Thermodynamics and Its Statistical Foundations (Oxford University Press, Oxford 1981 and 1983)
15. M.C. Tringides, P.K. Wu, and M.G. Lagally: This volume

Diffusion at the Vacuum-Solid and Gas-Solid Interface: Some Comments

J.H. Block

Fritz-Haber-Institut der Max-Planck-Gesellschaft,
Faradayweg 4–6, D-1000 Berlin-Dahlem (33), Germany

The vivid discussions during the first day of the Joint Workshop left no time for the "Summary by the Chairman" which had been announced in the program. The present comments are therefore, intended to summarize some of the major directions of these endeavors. However since the general discussion was held later in the day, the disputes that arose there will be advantageously incorporated. The contributed papers of the evening will also be considered.

Reactants are brought into physical proximity by surface or interface diffusion. This is a necessary condition for crystal growth, cluster formation or chemical reactions. In many of these actions diffusion determines the time scale of the over-all process. It is this behaviour which makes diffusion at interfaces of general importance for all kinetic events at solids. There was no need to emphasize this importance during the workshop. However, it turned out that the phenomenological treatment as well as the theoretical - not even microscopic - concepts of interface diffusion require major consideration and development. This applies to the macroscopic as well as the microscopic levels. One of the apparently simple questions, whether 'hopping' of surface particles may explain 'diffusion', caused an extended dispute. H.J. Kreuzer distinguished between diffusion as macroscopic mass transport (Fick's law) and hopping as a Markov process /1/. Only at conditions close to equilibrium are these two phenomena comparable, otherwise nonlinear phenomena will be important and require separate treatment. In phenomenology, this differs slightly from definitions of a "tracer diffusion coefficient" or a "chemical diffusion coefficient"as used for instance by Gomer /2/.

Both invited lectures of the morning session had high scientific level and excellent presentation giving the referee an easy task for his comments. S. George directly discussed surface diffusion in the new experimental method of laser pulse desorption, whereas P. Pukite had to consider surface diffusion indirectly as one limiting case during investigations on molecular beam epitaxy.

Since there have been only a few reliable macroscopic methods for investigating surface diffusion (like Auger work function, etc.), the laser-induced thermal desorption (LITD) technique of Viswanathan /3/ and George /4/ has been accepted as a valuable addition. It has the potential to be combined with standard surface characterization techniques, such that well defined surfaces of a great variety of solid materials promisingly may be analyzed. The experimental results of this new elegant method, concerning H_2-adsorption on Ru (0001), are impressive. One advantage of these systems is that diffusion coefficients on clean surfaces reveal data, which are coverage independent over a wide range of coverages. This facilitates data evaluation considerably. As discussed in the literature /2/, the determination of diffusion coefficients by concentration profile evolution is a straight forward analysis, if intermolecular forces - i. e. concentration dependent phenomena - are absent. Otherwise, the entangled Boltzmann-Matano procedure has to be applied.

The detection of only classical isotope effects for H_2 and D_2 at $T > 260$ K is actually not unexpected for the referee, since tunneling contributions have only been observed at $T < 120$ K with other methods. The models derived from experiments with coadsorbed carbon, sulfur or carbon monoxide are convincing. With these three different substances on Ru (0001), three different mechanisms of adsorbate interaction on hydrogen diffusion are derived. Such investigations are important, since they provide a link between "pure surface diffusion" and "mixed surface diffusion", which frequently governs surface phenomena like heterogeneous catalysis. The diversified influences of coadsorbed species shed light on the complications which those processes may envisage.

Systems such as Ni (001) or Ru (0001) and hydrogen are obviously good choices for demonstrating the abilities of the new method. This will, however, not exclude possible complications for other systems. There are some experiences from other related methods, like pulsed laser field desorption /5/ or reactive thermal desorption /6/, which may be taken to predict the general applicability or inapplicability of LITD.

Of main concern are molecular processes during the surface heating within the ≈ 300 µm spot illuminated by the ≈ 100 to 1000 nsec long laser pulse. Reproducible results can be expected only if after each laser shot the identical, clean surface with the same morphology (step density) is restored. Any kind of surface deposit as observed for many systems /5, 6/, due to chemical reaction or dissociation, will interfere.

Another limitation of the method is associated with the fact that the activation energy of diffusion and the heat of desorption of most systems are not independent entities. The measurement requires that the desorption rate during the time (≈ 100 sec) of profile evolution is negligible. Otherwise the diffusion coefficient has to be sufficiently high to refill the desorption area. These requirements will be found only in limited temperature regions for many systems.

We have to keep in mind that LITD is a macroscopic method. Microscopic models derived in this paper are based on outside information. Nevertheless, interesting further contributions have to be expected from this promising technique.

Applications of LITD have been reported in the evening section in a contributed paper by E.G. Seebauer and L.D. Schmidt. It was, indeed, most interesting to learn about the large scale of diffusivities, about the differences which obviously exist in similar systems. Who can explain that D_o for CO / Pt (111) is only 10^{-2} of D_o for CO / Rh (111), two systems which otherwise behave similarily?

Pukite's invited talk showed how the elaborate and intensive application of the long known RHEED technique - which has not found as much attention as LEED - leads to remarkable results The advantage of RHEED is utilized when completely flat surfaces are available which is necessary for obtaining ideal streak patterns. The RHEED pattern then gives an overall view of the reciprocal lattice in one given azimuth. Broadening and splitting of the streak pattern serve for characterizing the surface morphology, i. e. an ordered array of equidistant steps causes sharp splitting with a separation inversely related to the terrasse length and broadening indicates disorder in the terrasse length or in the sequence of up- and down-steps.

In earlier work, the layer by layer growth on reconstructed or unreconstructed surfaces of semiconductors (Ge, GaAs etc.) was studied. In the first case, orbital pairing or bond rehybridization of reconstructed surfaces require an activation energy (bond breaking) in order to form the nucleus for a new layer. A step propagation mechanism (Burton-Cabrera-Frank-mechanism) is likely if surface diffusion is sufficiently high. A competing reaction is the formation of various clusters which finally combine to a monolayer. This clustering mechanism is mainly investigated in Pukite's contribution.

The growth on GaAs and Ge (100) layers is investigated under different growth conditions, with the aim to demonstrate another powerful capability of RHEED, namely that of determining cluster shapes from beam profiles. If clusters are formed on terrasses, the RHEED beam intensity reflects the monolayer perfection. The observed intensity follows the cyclic behaviour of monolayer period, and RHEED intensity oscillations appear. By adjusting temperature and growth rate, those temperatures can be determined where oscillations vanish. From these data and the knowledge of terrasse size, diffusion coefficients are evaluated. The results of Pukite's work are striking in their details: concerning anisotropy of surface diffusion, shape of clusters, reactivity of different crystal sites and influences of impurities. One should, of course, keep two aspects in mind: (i) RHEED does - like LEED - neglect disordered areas of the surface and the real surface therefore may drastically differ from the idealized ones and (ii) all the considerations are based on models and conclusions apply within these models only. Flanking experiments with other methods - for instance scanning tunneling microscopy - are strongly demanded. Considering the great technical importance of molecular beam epitaxy, investigations like those of Pukite will find increasing attention.

Two further interesting contributed papers were presented in the evening: M. Karimi and G. Vidali reported about their model calculations on rare gas mobilities on the (001) MgO surface and compared these results with data on graphite. They could calculate the corrugation of the potential going from the most to the least adsorptive site. These results demonstrate a valuable contribution to decribe potentials in front of a solid surface.

The paper by M.C. Tringides, P.K. Wu and M.G. Lagally leads back to a problem mentioned at the beginning of these comments The diffusion behaviour at equilibrium and far from thermodynamic equilibrium. Investigations on the scaling in the ordering kinetics of oxygen on W(110) are performed with great precision, and profound theoretical models are used for data evaluation. The remarkable results of differences in diffusion at equilibrium and non-equilibrium conditions prove in a much more detailed quantitative manner those ideas, which have long been discussed in heterogeneous catalysis, for instance in 1931 in G.M. Schwab's book /7/.

At the small, inspiring workshop at Campobello only a selected series of problems in surface diffusion could be considered. These have been subject to vivid discussions. A more general overview of the field will probably be published in the proceedings of the Nato Summer School in Feldhofen /8/.

/1/ H.J. Kreuzer, this volume

/2/ R. Gomer in "Surface Mobilities on Solid Materials" Plenum Press 1983, p.1 ff, V.T. Binh editor

/3/ R. Viswanathan, D.R. Burgess jr. , P C. Stair and E. Weitz. J. Vac. Sci. Technol. 20 (1982) 605

/4/ S.M. George, M. DeSantolo, R. B. Hall, Surf. Sci. 159 (1985) L 425

/5/ W. Drachsel, S. Nishigaki and J.H. Block, Int. J. Mass Spectrom. Ion Phys. 32 (1980) 333

/6/ R.J. Madix, Chemistry and Physics of Solid Surfaces II, R. Vanselow ed. CRC Press 1979, p. 63

/7/ G.M. Schwab "Katalyse vom Standpunkt der chemischen Kinetik", Springer-Verlag Berlin 1931, p. 13 ff

/8/ P.K. Larson in Nato Advanced Study Institute, Plenum Press New York in press

Part II

Diffusion at the Solid-Solid Interface, Surface Segregation and Bulk Diffusion

Diffusion at Surfaces: Introductory Remarks

E. Zeitler

Fritz-Haber-Institut der Max-Planck-Gesellschaft,
Faradayweg 4–6, D-1000 Berlin-Dahlem (33), Germany

Möge es bald einem Forscher gelingen, die hier aufgeworfene für die Theorie der Wärme wichtige Frage zu entscheiden! (*A. Einstein*)

As a motto for this introduction to the chapter on surface diffusion, I have chosen the very last sentence of Einstein's famous and incisive paper [1] in which he relates the coefficient of diffusion to the (root mean square) displacement of a particle and the time elapsed between the initial and the final observation. Even he, Albert Einstein, asks for experimental confirmation of his deductions. Since the three contributions here are experimental observations and measurements, it seems that Einstein's hope is being fulfilled.

The three papers deal with the transport of matter, say atoms, over surfaces; but they differ in the way this transport is determined and also in the distances over which the transport takes place.

But before discussing details, let me set the stage properly by going back to Einstein, who shows that in a statistical random walk (drunken sailor) the diffusion coefficient is

$$D = 1/2\ \Gamma \ell^2$$

where Γ is the frequency of uncorrelated jumps and ℓ is the distance covered by a jump. (The factor 1/2 pertains to a linear migration; 1/4 would be for a two-dimensional problem; and in three dimensions it depends on the crystal structure, but 1/6 would be correct for a cubic crystal.) A typical value [2] of D for surface diffusion of Ag on Cu(111) at 635 K is

$$D = 10^{-7}\ (s^{-1}\ cm^2).$$

We get a better feeling when we express the distances in terms of intervals between adjacent sites on the surface of the lattice. Then

$$D = 1.6 \cdot 10^8\ (s^{-1})\ (2.5\ \text{Å})^2.$$

Is $\sim 10^8\ s^{-1}$ a high frequency? The question gives me the opportunity to introduce a very important name which does not occur in the papers to follow. If you have thermal oscillations about an equilibrium spot, a good estimate for the frequencies is given by the Debye frequency via

$$\Gamma_D = k/h\theta = 2 \cdot 10^{10}\ (s^{-1}\ K^{-1})\theta.$$

As the Debye temperature θ is of the order of a few 100 K, Γ_D amounts in the case of Cu to $6 \cdot 10^{10}$ Hz, which is very large compared to the jump frequency of diffusing particles, so that this diffusion appears as a slow-motion process in the surrounding turmoil of the vibrating captives. To complete the scenario, let us estimate the mean square amplitudes of the thermal vibrations as they are known from structural research and which are connected with Debye's and Waller's names. Near the Debye temperature we find for the mean square amplitude

$$\langle \Delta x^2 \rangle \langle 2Mk\theta \rangle = \hbar^2 (6T/\theta).$$

At room temperature the root mean square amplitude for Cu is .145 Å.

The low success rate for a wildly vibrating atom to escape to neighboring sites has two reasons: First, the moves bring about a change in the configuration of the assembly. As some configurations are more likely than others, the corresponding changes in entropy ΔS must be considered. The second reason is the energy barrier which a diffusing atom must overcome in order to leave its cell. Tunneling through is impossible on account of its heavy mass M. In the example above, the activation energy necessary to overcome the energy barrier is 17.5 kcal per mol, or 0.76 eV per atom. The Boltzmann factor, which predicts the likelihood of extracting this adequate amount of energy from the thermal reservoir, is $\exp(-13.15) \sim 2 \cdot 10^{-6}$, whereas $\Gamma/\Gamma_D = 5 \cdot 10^{-5}$. These two values differ by a factor of 25. If this factor is ascribed to the configurational entropy change, a value for ΔS of 3.2 k results, which is very reasonable [3]. This agreement is almost too good for such coarse reasoning, but it reflects the essentials. When the height of the potential barrier is great compared to the thermal energy, the problem can be treated as quasi-stationary. Its treatment by Markoff's method, as well as deep insights into the general stochastic process in physics, can be found in a most scholarly treatise by S. Chandrasekhar [4].

And now to the experiments. Each invited paper has a single author, hence it should be simplest to refer to the paper by the name of that author. Since diffusion is a statistical process, all three authors must come up with measured averages. None of the instruments permits direct observation of the elementary act, the jump. Therefore, the moving atoms must be given a certain time interval in which to move over the distances or produce spatial changes that can be resolved by the instrument employed.

We begin with Fink's work because he uses the most basic method, following the elementary diffusion process as outlined above, in a field ion microscope capable of depicting single atoms which form a perfect plane at the top of a pointed tungsten tip. This plane has a diameter of about 60 Å or, better, of about 20 atomic sites. The strong electric field is cleverly used for preparing and surveying this unique atomic area. Onto this small "trampoline" adatoms are deposited whose manner of jumping is the subject of interest. Then the trampoline is heated up to 350 K, and over a period of 20 seconds the adatoms can, unobserved, do their dance. Thereafter the trampoline is cooled to below 50 K causing the atoms to sit still so that a photo of their position can be taken. After this, a further shake-up is initiated, because thousands of displacement data are needed for good statistics. On the other hand, the trampoline cannot be overcrowded because that would lead to confusion of the evaluation. The accuracy rises with the square of the labor. The simplicity and beauty of this approach, however, belies the requirement for highest skill and patience. The method is capable of detecting sites with increased binding energies, as exemplified by the high occupation numbers at the edge sites of the tungsten plane. The movement of the entire rhenium emsembles such as the widely spaced noncontiguous Re-pentamere will surely elicit much discussion.

Behm uses for his investigations a scanning tunnel microscope--but not to atomic resolution. He is interested in finding out how physical or chemical inhomogeneities of the surface affect the diffusion of single adatoms. The physical measurement, the sought statistical average, is the change in size (in number of atoms or Å) of monoatomic islands of Pt surfaces at elevated temperature. Here we speak of 700 - 1100 K and of reaction intervals of minutes. Behm knows from reactive surface chemistry how to produce flat islands. Their shrinkage or growth is taken as an indicator for the adhesive forces between the free adatoms and the coastal regions of that island. In this respect the islands correspond to Fink's binding edges. The islands can grow so long as single atoms are available for diffusion which then get stuck on account of the high binding energy. At low binding energy those atoms have a chance to leave the islands and diffuse freely until they find a new island. This produces a distribution of island sizes markedly different from that in the high-binding-energy case. Monte Carlo simulations confirm the existence of these two distributions. Chemically different adatoms such as the carbon left from initial ethylene reaction may stabilize the surface structure to such an extent that small clusters form instead of islands.

And finally Blakely's work. As pointed out several times, diffusion measurements require averaging. Blakely does it elegantly by optical means. An interference microscope is utilized which tells us already that the geometric extensions concerned are several microns. Therefore we are not surprised that the reaction times are of the order of days, although with temperatures up to 1400 K.

Making use of modern lithography methods, Blakely etches the specimen Al_2O_3. By this means he produces an optical sinusoidal phase grating: the period is 7.5 microns and the height of the corrugation amplitude is 170 nm. If this grating were made from wax (think of an old phonograph disk) it is easy to understand that heating it would smoothen the surface and fill out the grooves. The surface tension (of the melt) does it. The same transport of matter takes place in the case of the alumina grating. Microscopically the atoms move along the terraces, hang on to the ledges and so on, with the end effect that the corrugation amplitude decays exonentially in time. The 1/e time is inversely proportional to the diffusion coefficient. And it is this decay which is directly visible and measurable in the interference microscope. Always when a coherent phase-sensitive process is disturbed by a stochastic (thermal) influence, the decay of the amplitude follows an exponential law, e.g., Debye-Waller factor, or the friction damping of the harmonic oscillator.

The intent of this introduction is to show how three papers--all concerned with material transport on surfaces by diffusion--relate to each other in scale, and to point up the differences in methodology for assessing this transport. The finding out of the details, the kinetic and dynamic peculiarities, should make the papers themselves interesting to read.

Literature

1. A. Einstein: Ann. Phys. 17, 549 (1905)
2. C. A. Roulet: Surf. Sci. 36, 295 (1973)
3. P. Wynblatt: Phys. Stat. Sol. 36, 797 (1969)
4. S. Chandrasekhar: Rev. Mod. Phys. 15, 1 (1943)

Direct Observation of Atomic Motion on Surfaces *

H.-W. Fink

IBM Research Division, Zurich Research Laboratory, 8803 Rüschlikon, Switzerland

1 INTRODUCTION

Direct insight into the atomistics of the processes governing the diffusion of adsorbates on surfaces requires the ability to visualize the atoms carrying out the motion The field-ion microscope (FIM), invented by Erwin MULLER [1, 2] in 1951 is clearly capable of imaging surfaces with atomic resolution Quantitative studies of individual atomic processes on surfaces were first presented by EHRLICH and HUDDA [3] in 1966 with their findings of the self-diffusion of tungsten on different crystallographic planes of a tungsten tip With this, an era of field-ion work began that has led to quite an extensive amount of information on surfaces at an atomic level [4, 5]. Here, only a fraction of this work will be presented in some detail, mainly the result of collaboration between Gert Ehrlich and the author during his stay at the University of Illinois at Urbana-Champaign. Apart from this subjective selection, the findings presented here belong to a group of experiments that demonstrate a complexity of atomic behavior not anticipated on the basis of traditional models of binding and interaction

First, a short description of the technique of FIM, with special emphasis on the procedures important for diffusion studies, is given. The presentation of experimental results starts with the findings on single-atom diffusion on flat surfaces. Throughout the paper, the systems presented will be of increasing complexity. The influence of lattice steps on single adsorbed atoms as well as the binding of atoms to overlayers are discussed in view of traditional models of binding A system in which the contributions of nonpairwise interactions are solely responsible for the cohesion of clusters is presented next Finally, the diffusion properties of clusters exhibiting unexpectedly high mobility are illustrated

2. TECHNICAL ASPECTS

The image in the FIM is formed by the ionization of neutral image gas atoms in the presence of a high electric field of the order of 4 V/Å. A sharp tip, as suggested in Fig. 1, is put at some kV high potential in respect to an ion detector, usually a microchannel plate assembly. The sharply pointed tip produces a highly inhomogeneous field at the tip surface that attracts image

* Work performed at the Coordinated Science Laboratory of the University of Illinois at Urbana-Champaign under NSF grant DMR 82-01884

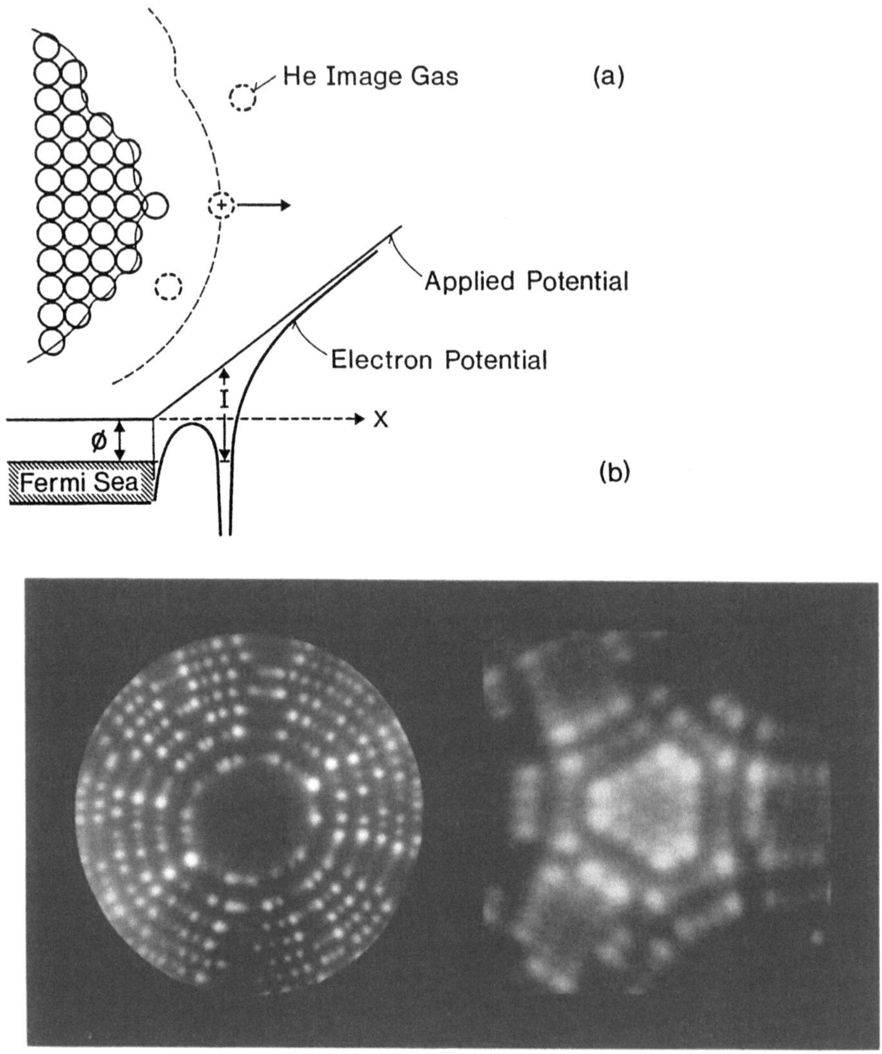

Fig 1 Field ionization of helium at a metal tip surface. (a) Cross section of tip with dashed line indicating the critical distance which has to be exceeded for ionization to occur. (b) Electron potential at distance x from the surface I = ionization potential of free atom; ϕ = work function of the metal surface. Bottom: Field-ion pattern of a (110) at left, and a (111)-oriented tungsten tip at right

gas atoms owing to their field-induced dipole moment Eventually, they become ionized at or beyond a critical distance by tunneling of a valence electron into states of the tip surface at or above the Fermi level The barrier for this tunneling process is given by the superposition of the applied potential onto the atomic potential for the electron in the neutral gas As a consequence, the positively charged ions are accelerated along the electric-field lines towards the detector where they form the image spots. The various models and experimental findings on the details of the ionization processes themselves can be found elsewere [6]

For the scope of this paper, it is sufficient to note that the field-ion pattern is a highly magnified projection of those locations on the tip surface, where the local protrusion and consequently the electric field is high This determines the tunneling probability for the electron and with it the ionization rate for the noble gas atoms Apart from the electric field, image contrast may also be influenced by the supply of certain sites with image gas atoms and the local electronic structure of the tip surface In Fig 1, two field-ion micrographs of differently oriented tungsten tips are also shown With the (111) orientation of the tip, the variation of the electric field at the relatively open apex plane is sufficient to resolve the atoms within the plane, as evident from the micrograph shown at the right The (110) terrace on the other hand, shown at the left, is densely packed, therefore appearing dark in the field-ion microscope Only the steps to the vicinal planes, which constitute high local protrusions, are readily imaged

The fact that the surfaces under investigation in the FIM are relatively small facets of a tip brings with it the unique possibility of this technique to create perfect surfaces by field evaporation of surface atoms at low temperatures The evaporation starts at the edge of the plane where the electric field is highest. This allows preparation of the desired size of the plane within a given range The (110) plane shown in Fig. 1 was field evaporated to a diameter of roughly 60 Å

2.1 FIM Experiments on Diffusion and Atomic Interactions

There are some common procedures in all the designs of FIM experiments that aim at diffusion as well as interaction studies between atoms To set the stage for what follows, these similarities in the setup of such experiments are now discussed

After the cleaning procedures of the tip, usually achieved by heating and sputtering, have been performed, the apex plane is field evaporated to the size desired. This is followed by the deposition of adsorbates from a source facing the tip If more than the desired number of atoms have settled on the plane, the excess atoms will be removed by field desorption This can be done in a controlled manner one atom at a time since the adatoms closest to the edge of the plane experience the highest field and will therefore desorb first when the field is increased The ones closer to the center of the plane then become available for an experiment To study the displacement of the adatoms, the voltage is switched off while the tip is heated to a preset temperature for a certain period After this heating interval, a waiting time is imposed, as indicated in Fig 2 Only after the tip has reached a temperature

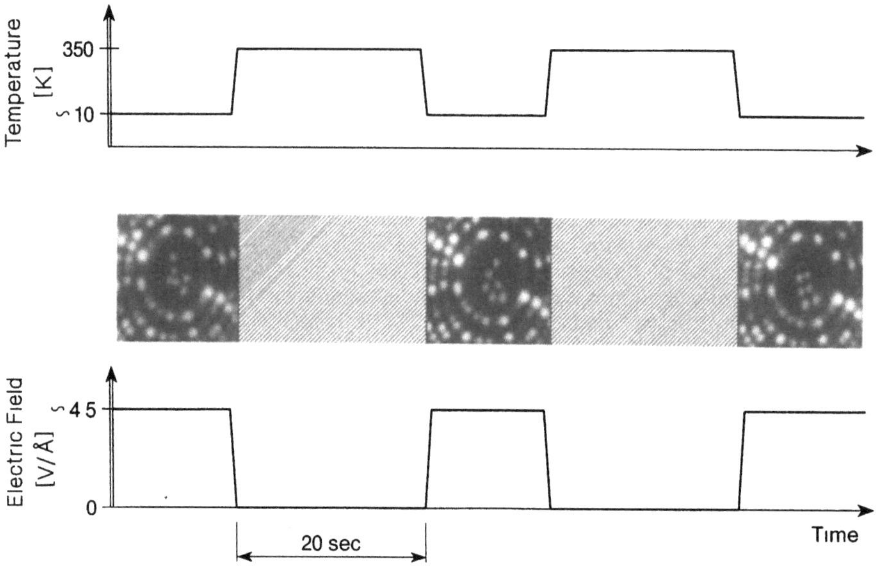

Fig. 2. Time sequences involved in an FIM diffusion study; six rhenium atoms have been deposited on the (110) face of tungsten The system is repeatedly heated to 350 K for 20 sec time intervals to achieve mobility of the atoms

below 50 K, is the field again applied to observe the new adatom configuration. In this way, it is assured that the motion of the atoms takes place in an ordinary thermal environment, not affected by the electric field.

In any new adsorbate system to be investigated, it is important to verify that the field alone does not cause motion of atoms at low temperature. This is checked by switching the potential at the tip on and off. In a modern FIM setup, heating and imaging are cycled automatically and the pictures taken with a low-light level video camera and stored on tape for later analysis [7].

3. DIFFUSION OF INDIVIDUAL ATOMS

The notion how atoms move over a crystal surface is indicated by the schematics in Fig. 3, it illustrates the elementary process in surface diffusion. Bound to a specific site on a surface, an adsorbed atom encounters a barrier in hopping to one of the energetically equivalent neighboring sites The barrier can be overcome by thermal activation mediated by collisions with the substrate atoms The jump rate is described by an attempt frequency Γ_0 and the barrier height E_d,

$$\Gamma = \Gamma_0 \exp(-E_d/kT) .$$

These two quantities were measured direct by EHRLICH and HUDDA in 1966 [3] with individual tungsten atoms on various crystal planes of tungsten. The data are obtained by observing the new adatom position after each heating

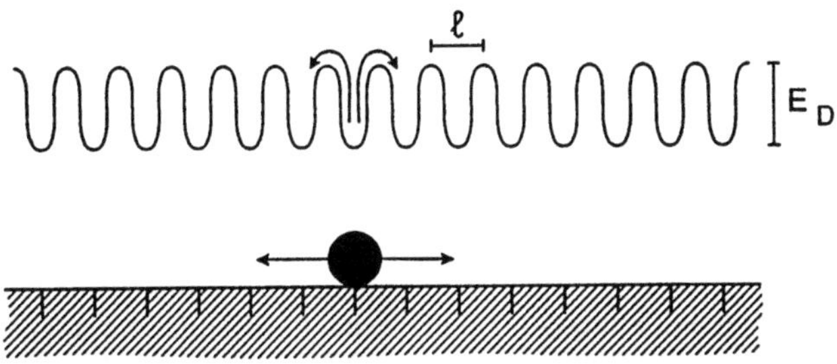

Fig. 3 Schematic diagram of the free energy for a single adatom undergoing displacements on a surface; ℓ = distance between adjacent sites, E_D = activation energy for diffusion

cycle of duration τ at a temperature T A long sequence of such observations reveals the mean-square displacement $\langle \Delta X^2 \rangle$ during an interval τ. For a random walk, $\langle \Delta X^2 \rangle$ is related to the jump rate Γ and the mean-square jump length ℓ^2 by

$$\langle \Delta X^2 \rangle = \Gamma \tau \ell^2 .$$

The diffusivity D in a one-dimensional diffusion process is tied to the mean-square displacement by Einstein's relationship

$$\langle \Delta X^2 \rangle = 2D\tau .$$

With $D = D_0 \exp(-E_d/kT)$, measurements of the temperature dependence of the mean-square displacement reveal both the barrier E_d and the prefactor D_0 which is the product of the attempt frequency Γ_0 and ℓ^2. For jumps with zero activation entropy at room temperature $\Gamma_0 = kT/h$, the prefactor should be of the order of 10^{-3} cm^2/sec

In this way, much has been learned about the mobility of individual metal atoms on various surfaces [4, 5] In general, the notions outlined above seem to describe the processes of diffusion quite well in most of the systems examined. It turns out that the surface structure plays an important role for the activation barrier This is most clearly shown in studies of self-diffusion on rhodium: On the close-packed (111) plane, diffusion already occurs at cryogenic temperatures, whereas on the rougher surfaces, atoms do not move until room temperature values have been reached [8]. The dynamics of motion, reflected in the prefactor D_0, also appears normal. However, there is one example where the experimental findings drastically deviate from the picture outlined above: For diffusion on some surfaces of the platinum-group metals On the channeled platinum (110) surface BASSETT and WEBBER [9] observed motion both along and across the channels, and on the Ir(110) WRIGLEY and EHRLICH found diffusion to occur mainly across the channels With the atom-probe field-ion microscope they were able to establish that, for W on Ir(110), diffusion takes place by an exchange mechanism [10] The adsorbed tungsten atom pushes an iridium lattice atom out of its normal position into a neighboring channel, and occupies its position in the lattice The iridium atom then continues the migration process This unusual diffusion mechanism might well be correlated to the reconstruction also observed for those noble-metal (110) surfaces For rhodium on the other hand, not known to exhibit a reconstructed (110) surface, self-diffusion occurs in a "normal" manner. This might be an indication that the behavior of individual atoms, being a probe for the surface potential, is also tied to the forces responsible for the reconstruction of surfaces

4 THE INFLUENCE OF STEPS ON SINGLE ADATOMS

The importance of lattice steps is realized in many aspects of surface science like atom scattering, chemisorption, catalysis and electron emission [11]. The question that should be addressed here is how the presence of a lattice step alters the binding of a metal atom: A matter of importance to understand transport and growth processes on a metal surface. Consider a single adatom on a line of sites that is determined at both ends by a descending step as indicated in the schematics at the bottom of Fig 4 There is a discrete number of adsorption sites available to the atom, and the questions are what is the free energy of binding of these sites and how is it influenced by the presence of the steps? In a constant temperature ensemble, the probability P_i

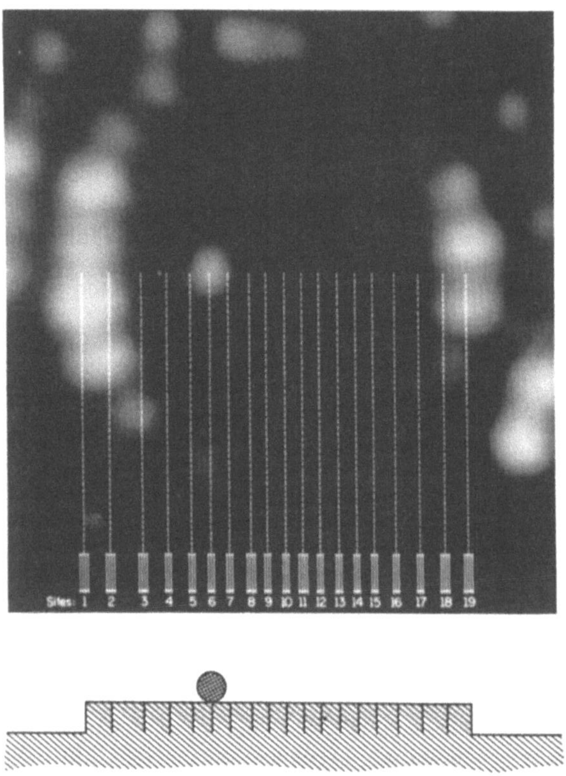

Fig 4 Single atom on a crystal terrace bounded by two descending steps. Top: Field-ion micrograph of a single W atom on a W(211) terrace on which 19 sites have been identified. Bottom: Schematic side view

of finding the atom in position i compared to the probability P_0 of finding it in a reference position unaffected by the edges reveals F_i, the free energy of the system with the atom at the various sites i,

$$P_i/P_0 = \exp(-F_i/kT)$$

For such an experiment, we selected the (211) plane of tungsten [7] On this surface, diffusion is one-dimensional along the [111] channels. The number of sites that have to be explored is therefore minimal. For a given number of diffusion cycles, this ensures a reasonable value for the occupation probability of each site that is not seriously affected by statistical fluctuations. To reach the highest possible number of observations on one and the same plane, we adopted an automated system to fulfill the necessary sequences of heating and observing. Individual tungsten as well as rhenium atoms were probed. A tungsten atom, deposited from the gas phase onto a (211) plane, the apex of a tungsten tip, is shown in Fig. 4. The channels in this picture are along the horizontal direction, and the central one in which the atom is adsorbed offers 19 sites on this terrace

To obtain an equilibrium distribution in which the occupation probability is only reflected by the energetics and not the dynamics of motion, Monte-Carlo simulations to determine the appropriate heating interval lengths and temperatures have been performed [7] High temperatures seem desirable because they keep the heating intervals short and allow a maximal number of observations in a given time On the other hand, too high temperatures result in frequent loss of the adatom over the plane edge, which however is reflective for most metal atoms A compromise between temperature and heating time has to be found to obtain an optimal number of observations with one and the same atom on a given plane An equilibrium distribution for a single W atom in such a channel is shown in Fig. 5. A total number of 1340

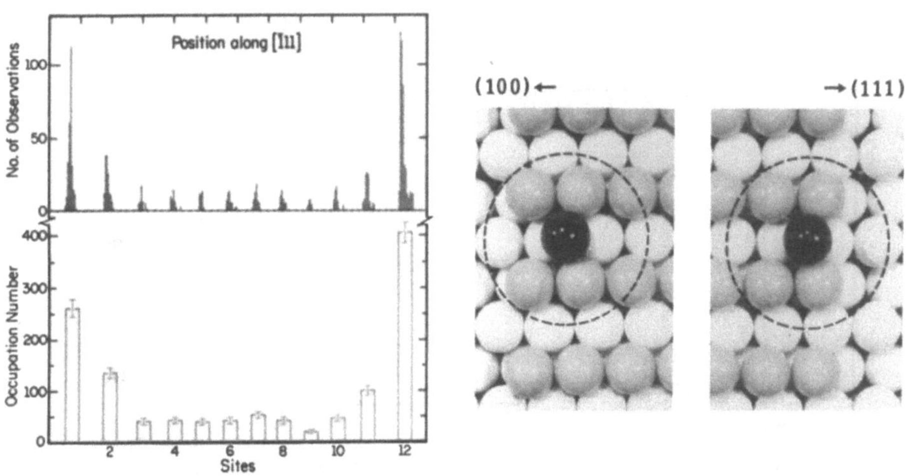

Fig 5. Equilibrium distribution of a single W atom on a (211) plane at 363 K. Left: Frequency of observing adatom locations; 12 sites can be clearly distinguished The lower graph shows the distribution of the occupation over the 12 sites based on 1340 observations. Right Hard-sphere model showing the differences in the geometry of the two edge positions

observations was made here to probe the 12 sites available to the single atom; the equilibration was performed at 363 K for 5 sec intervals. A small dispersion is apparent in the atomic positions during subsequent occupation of one and the same site by the adatom. This reflects the accuracy in the determination of atomic positions. The spread is small enough not to cause any problems in assigning the atom to individual binding sites on the surface which reveals the distribution shown on the lower left of Fig. 5. This distribution is far from uniform, and it is immediately apparent that the two edges bind the W atom more strongly than the sites on the flat terrace.

In Fig. 6, the relative free energies of the W atom in the 12 sites are shown together with a sample taken on a larger (211) plane with 23 sites and a total of 989 observations. The nonuniformity of the binding is reproducible from one sample to the other It is apparent that at the two edges of the (211) terrace the binding is different. The one facing towards (111) binds the W atom rather more strongly than the one in the (100) direction. This asymmetry could have been expected since on the (211) plane of a bcc crystal there is not a plane of symmetry perpendicular to the diffusion direction along the channels, as can be seen from the ball model in Fig. 5. However, the edge facing (111) offers fewer nearest neighbors than the step position facing (100); yet it binds the W atom more strongly.

Not even the qualitative aspects of the binding at steps can be explained by models based on pairwise additive interactions via a monotonic potential between atoms. In such a simple model, binding would be strongest in the

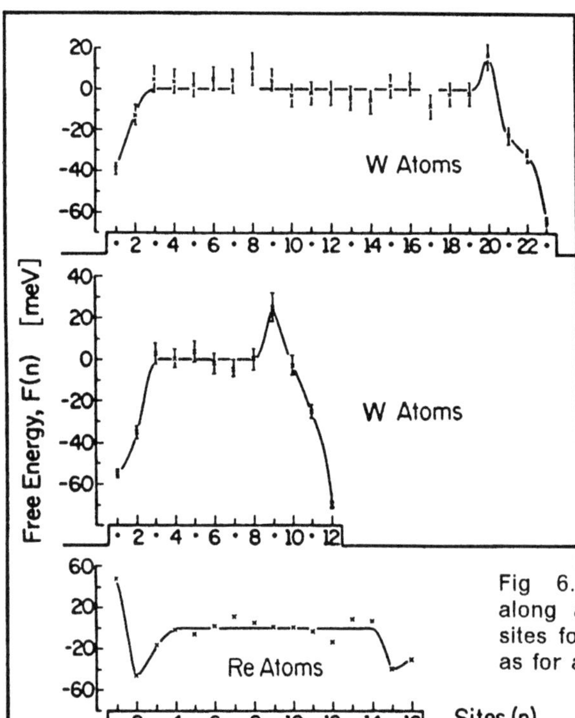

Fig 6. Free energy of binding along a terrace with 23 and 12 sites for a single W atom as well as for a Re atom over 16 sites

most highly coordinated sites, i.e. on the flat terrace Instead, the experiments exhibit a contrary behavior The influence of the step is quite long range, on the (111) site it even affects binding up to the fourth site from the step where a siqnificant increase in the free energy is noticeable

By equivalent experiments substituting the W atom by a rhenium one, the binding for a chemically different atom was examined With this it became evident that these edge effects are strongly dependent on the chemical nature of the adatom The distribution for a Re atom over 16 sites, also presented in Fig. 6, shows that the binding does not increase as sharply as for W in moving from the central sites towards the edges Not the outermost positions on the terrace are most attractive as for self-adsorbed atoms; Re seems to prefer the sites second nearest to the edge

It is not yet clear what causes these effects, but it seems likely that at the edges a redistribution of the lattice atom positions takes place that might be associated with a changed electronic structure. This is probed by the chemisorbed atoms and accounts for the rather unexpectedly rich behavior that is not at all congruent with simple models of binding on surfaces.

5. ATOM BINDING AT OVERLAYERS: Re ON W(211)

The system Re on W(211) has been studied quite extensively; the diffusion properties of individual atoms as well as dimers and trimers have been investigated quantitatively [12]. This provides a solid basis for investigations of more complex systems in which many atoms are involved.

We now concentrate on the interaction of single Re atoms with pseudomorphic-grown Re overlayers on the (211) plane of tungsten. This problem has some similarities with the single-atom step interaction presented in the previous section. An individual atom now interacts with the lower part of a step formed by the edge of an overlayer In this system, the pair interactions are well-known from studies of Re dimers [12] By subsequent deposition of Re and heating to above 500 K, a coherent Re overlayer can be formed on the (211) plane like the one shown in Fig 7. The structures of these layers have been found to be pseudomorphic with the underlaying surface [13] The deposition of one additional Re atom in a channel adjacent to the layer results in the incorporation of the atom in the layer; it is bound to its edge and can occupy varies edge positions, as shown in Fig. 7. Dissociation of atoms attached to the edge of these layers was not observed at 379 K. From the time the atom spends in the end position, we can derive a minimum lifetime and with this estimate a binding energy of at least 0.3 eV with which the atom is attached to the layer. Quantitative experiments for the relative binding to these layers for atoms in adjacent channels have been performed for four and five available sites next to a Re layer [13]. The result for a layer offering five sites is shown in Fig 8 The energies are derived from the probability with which sites are occupied under equilibrium conditions, analyzed in a way simular to that discussed in the previous section. The relatively low binding at the end sites can be well understood in terms of the classical picture of binding at steps [14]. The asymmetry between the two ends is attributed to the (211) substrate structure which does not have a plane of symmetry perpendicular to the channel direction. What cannot be understood in view of classical models of binding is that the central site is significantly less likely to be occupied by the Re atom Why is the binding in

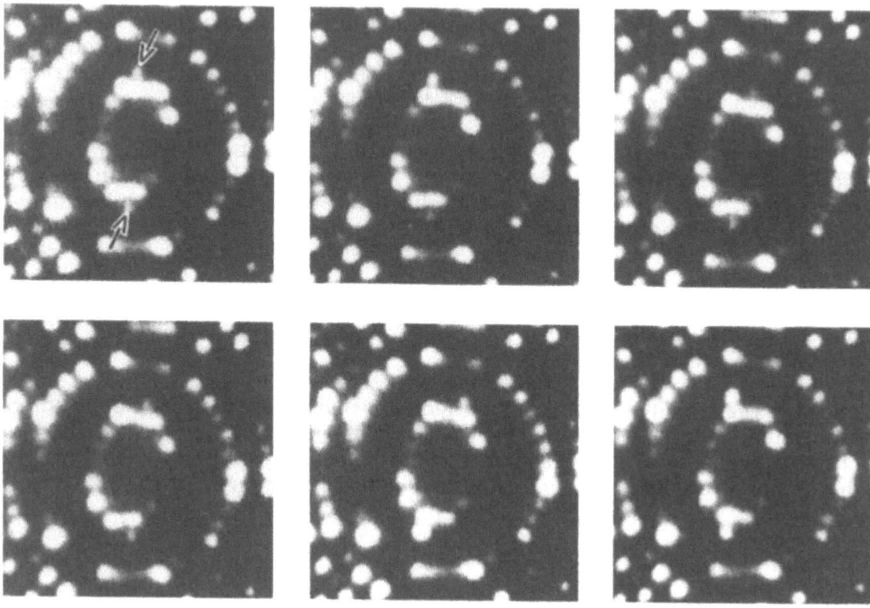

Fig. 7. Rhenium overlayer on W(211) with a single Re atom in the (111) channels adjacent to the layer. At temperatures exceeding 320 K, the individual atoms at both edges of the layer start moving along the layer edges

the central site, where the coordination of the atom is highest, lower than in the adjacent sites?

6. NONPAIRWISE INTERACTIONS BETWEEN ADATOMS

Instead of giving an answer to the question posed above (I do not know one anyway!) I should now like to discuss an experiment where the concept of pairwise interactions between atoms on a surface is not only a bad approximation but fails completely in describing the experimental facts; the existence of clusters and coherent overlayers [15].

6.1 Interaction Between Two Re Atoms on W(110)

The techniques outlined above are neither limited to single atoms nor to planes where diffusion is one-dimensional. Figure 9 shows the site mapping made with a single Re atom on a W(110) plane where two-dimensional motion of atoms is observed. Again, it is apparent that it is possible to distinguish between individual adsorption sites with the closest separation of 2 74 Å on this plane. This is a prerequisite to measure distances between adsorbed atoms.

Fig. 9. Mapping of the binding sites for Re atoms on the (110) plane of tungsten. Single Re atom at the left is allowed to occupy some of the sites several times during its random walk A close-up of an area of the plane with the sites actually observed is shown at the right

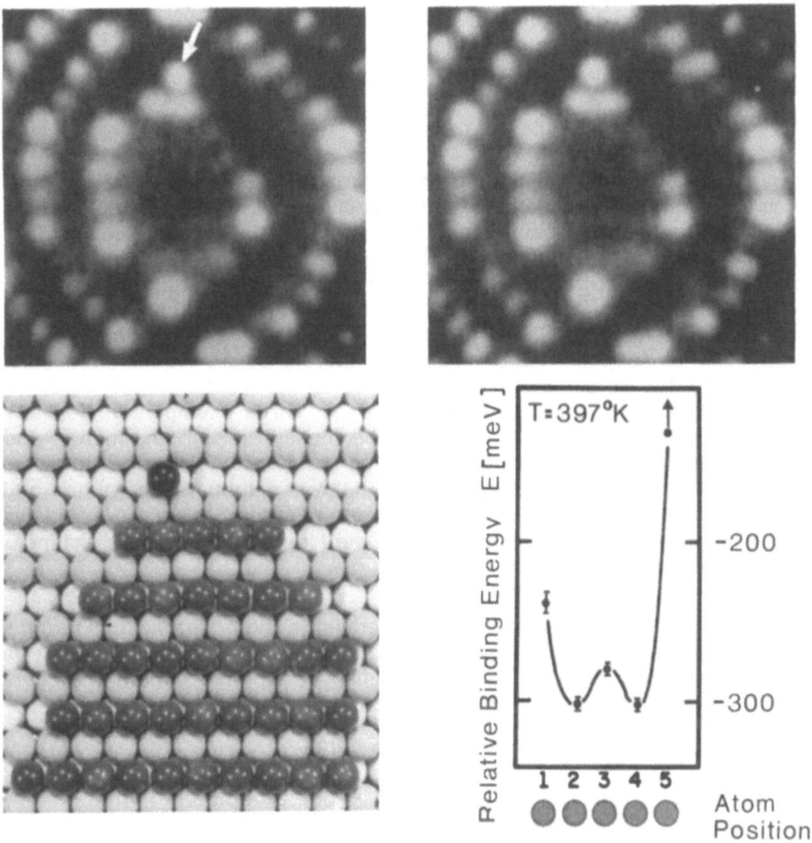

Fig. 8. Top: Re layer offering five sites at the top edge; the field-ion micrographs show the Re atom in two of the five possible sites. Bottom: Ball model on the left; relative free energy of Re atom in the five possible sites with the lowest free energy arbitrarily set to −300 mV

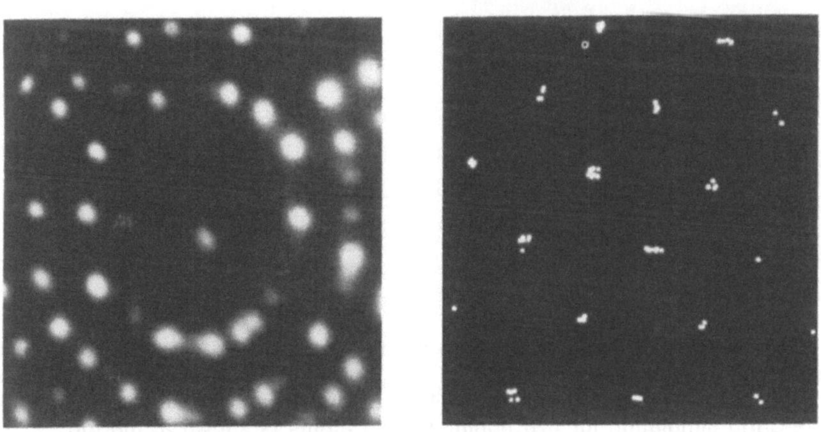

Fig.9. Caption see opposite page

The probability P(r) of finding two atoms at a specific separation r in a constant temperature ensemble is the quantity to give the free energy of interaction F(r) between pairs of atoms:

$$P(r) = CP_0(r)\exp[-F(r)/kT] \ ;$$

here $P_0(r)$ is the probability of finding the interatomic separation r when there are no interactions present other than prevention of double occupancy of sites; C ensures normalization. Equilibration in such a system means that between observations the atoms must be able to move freely over the sites available on the plane to feel each other's presence. These conditions as well as the pair distribution P_0 for noninteracting pairs have been derived by Monte-Carlo simulations [16]. In a total of about 1000 observations on Re pairs, only two close pairs were observed that dissociate immediately once the temperature for single atom diffusion has been reached This confirms previous studies that Re dimers are unstable [5] Three examples of Re pairs on W(110) are shown in Fig. 10 as well as the measured pair distribution and the one for noninteracting pairs. By comparing the two, it is apparent that the population of Re pairs at distances between nearest-neighbor separations and about 10 Å is siqnificantly below that for noninteracting pairs. A repulsive potential with an average value of about 50 meV operating at short distances can be derived from this.

What are the consequences of this repulsive pair interaction for the formation of overlayers in this system? If the binding in layers could be described by summing up over all pair contributions, then coherent clusters and layers of Re on the (110) face of tungsten should not be existent.

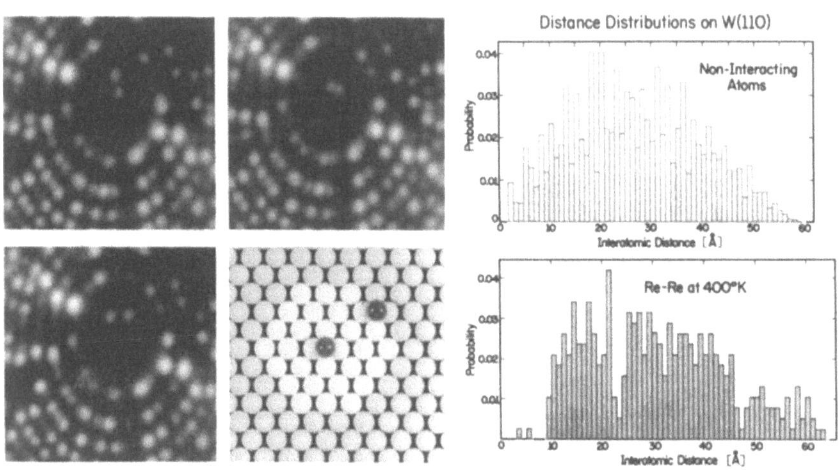

Fig. 10. Left: Re pair configurations on W(110), the atoms are seldom observed in spacings closer than indicated in the hard-sphere model. Right: Radial distribution for two Re atoms on W(110) at 400 K shown at the bottom. A Monte-Carlo simulation for two noninteracting atoms is shown at the top

6.2 Many-Body Interaction

The experiment shows that despite this repulsive pair interaction, Re overlayers are formed at higher coverage The cohesion in these overlayers must therefore arise from many-body interactions The smallest stable clusters observed are Re trimers, shown in Fig 11 Their pure existence is a clear indication that a nonpairwise trio interaction in the system Re on W(110) is present Studies on the lifetime of these trimers as well as experiments in which the equilibrium between trimers and individual atoms has been measured give a quantitative value of the cohesive energy of the trimer. It amounts to about 0.25 eV entirely due to that nonpairwise many-atom interaction [15, 16].

This particular behavior of Re on W(110) definitely cannot be generalized, but it clearly shows the importance of many-body interactions in chemisorbed layers that might however be much less pronounced in other systems. For a general approach to an understanding of cohesion in adlayers, it seems necessary to consider the presence of higher-order interaction. Theoretical efforts have already been made to explicitly calculate trio-interactions; a step towards arriving at more realistic models of binding between adsorbed atoms [17, 18]

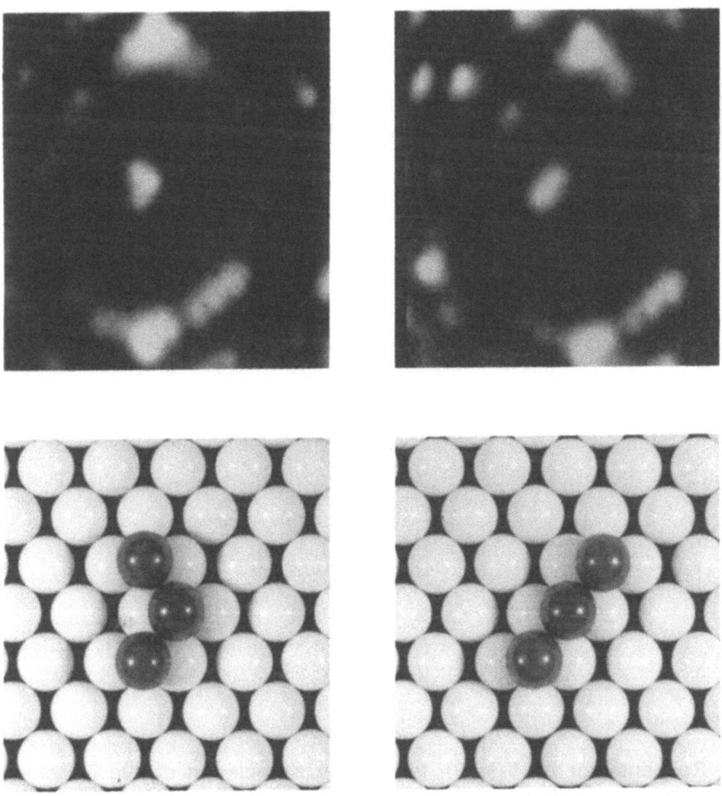

Fig 11. Re trimer configurations. Top: Field-ion micrographs, bottom: Corresponding hard-sphere models

7. DIFFUSION OF CLUSTERS

The Re trimers shown in Fig. 11 are found to be close-packed, with binding length corresponding to the 2.74 Å lattice atom distances along the close-packed directions of the (110) plane. During diffusion, the trimer changes from the straight to the triangular-shaped configuration. This is in agreement with a cluster motion taking place by uncorrelated single-atom jumps, which has been quantitatively investigated for small Re clusters on the (211) face by EHRLICH and STOLT [12] as well as for Pt clusters on the (110) W plane by BASSETT [5]. However, quite a different mechanism seems to be responsible

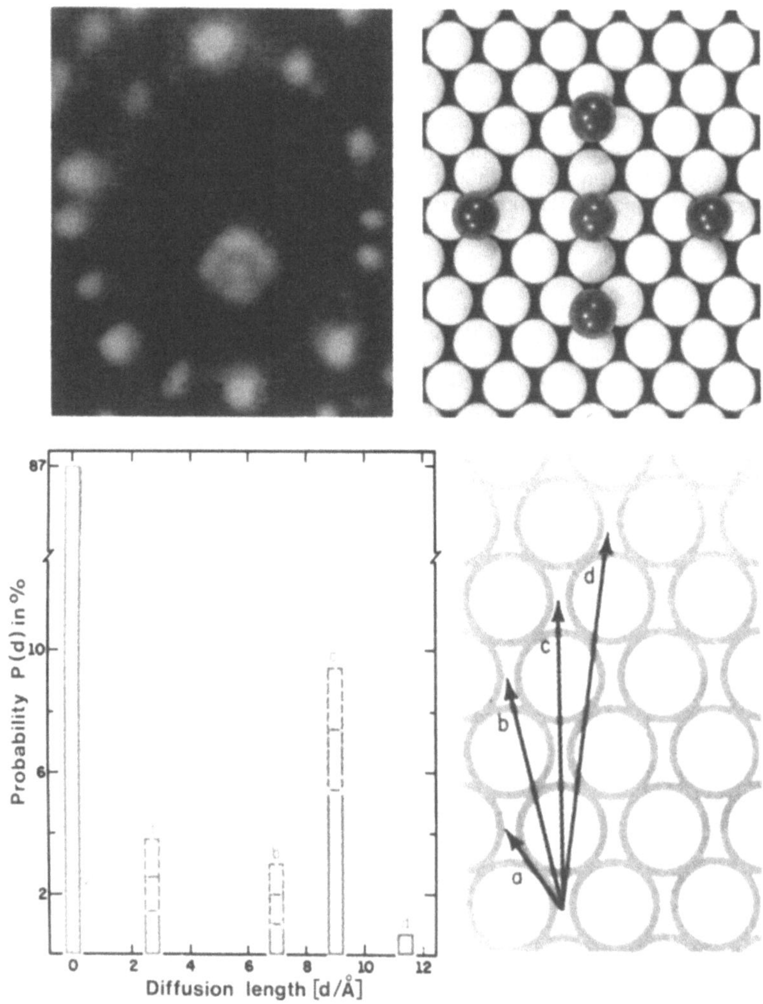

Fig. 12 Top: Re pentamer on W(110), formed by association of five Re atoms upon heating at 400 K. Bottom: Distribution of the displacement of the center of mass of a Re pentamer during diffusion over the W(110) plane at 336 K after 30 sec heating intervals

for the motion of slightly larger Re clusters made up of four or five atoms. A Re pentamer is shown in Fig. 12. In contrast to the trimer it has an open atomic arrangement as indicated in the ball model. Together with the open structure comes an unexpectedly high mobility compared to the Re trimer. The mean-square displacement of the center of mass of this cluster is even larger than that for a single Re atom for a given time interval and temperature [19]. For clusters that move by single-atom jumps, the barrier increases with the number of atoms contained in the cluster Such high mobility as observed for the pentamer is consequently hard to fit into this concept. The displacement distribution of the cluster has also been measured and is illustrated in Fig. 12. The most likely jumps occur along the (110) direction to second-nearest adsorption sites. Again, for motion by single-atom jumps, displacements to positions further than nearest-neighbor sites are unlikely

All this suggests a picture in which the pentamer moves over the surface by collective jumps of all five atoms at the same time. Owing to the binding among them and the interaction with the substrate, its motion might consequently be better described by a single barrier for the pentamer as a whole rather than by individual barriers for each of the five Re atoms.

Finally, one example for even larger aggregates of metal atoms that are still capable of moving as a compact unit over a surface is presented Growth and movement of clusters of Pd atoms on the W(110) surface are illustrated in Fig 13. It was first noticed by BASSETT [5] that the platinum metals grow in quite a complex manner on the W(110) surface. At low coverage they form

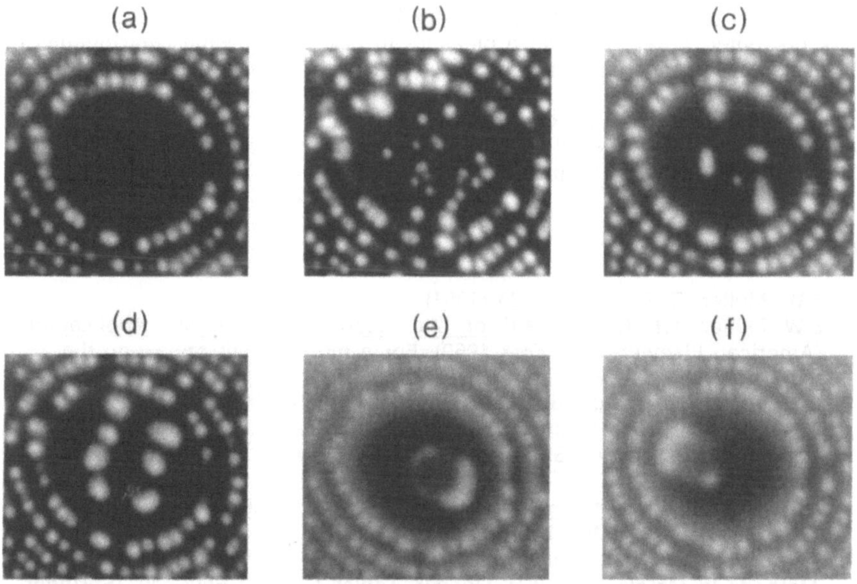

Fig 13. Formation and diffusion of Pd islands on the W(110) plane. (a) Clean W(110) surface; (b) deposition of roughly 20 Pd atoms, (c) after heating at 235 K Pd chains have formed; (d) after further deposition of Pd and heating at 235 K, two-dimensional clusters form; (e) clusters have associated to a large Pd island at 340 K; (f) the island moves over the surface at 390 K

one-dimensional chains oriented along the close-packed [111] direction, as shown in Fig. 13(c) for palladium. At higher concentration, two-dimensional clusters form that associate to quite large Pd islands containing to the order of 50 atoms as apparent from Fig. 13(e). What seems striking is that these quite large structures made up of many Pd atoms are not stationary but capable of moving over the (110) surface at barely 100° above room temperature.

8. SUMMARY

Although the findings presented here are only a very limited fraction of the data obtained by field-ion techniques over the last three decades, they demonstrate the capability of this technique to obtain detailed insight into the atomistics on surfaces. The behavior of adsorbed atoms in view of their dynamics and kinetics seems rich; challenging for the experimentalist to discover unknown facts on the one hand, while on the other hand, handy models and convenient atomistic pictures have been established that have only seldom been checked by experiments. Some of these models have definitely to be re-examined or modified to match reality. With better experimental insight, there will be the hope of finding more realistic models. They might even be simpler than the concept of pairwise interaction or bonding models relying on geometric atom coordinations.

ACKNOWLEDGMENTS

I am grateful to Professor Gert Ehrlich for his support during the course of the work presented here. It is also a pleasure to acknowledge the contributions of the glassblowing, photographic and machine shops of the Coordinated Science Laboratory at the University of Illinois. I am much indebted to our publications department at the IBM Zurich Research Laboratory for invaluable assistance with the manuscript. Special thanks I owe to Mrs. Helen Bernet and Mr. Urs Bitterli for their efficient help in the stages of preparing the final manuscript.

REFERENCES

1. E.W. Müller Z. Phys. 131, 136 (1951)
2 E.W. Müller, T.T. Tsong· Field Ion Microscopy, Principles and Application (American Elsevier, New York 1969). For a more recent review on the technique, see J A Panitz: J. Phys. E15, 1281 (1982)
3. G. Ehrlich, F.G. Hudda: J. Chem. Phys. 44, 1039 (1966)
4. G. Ehrlich: In Proc. 9th Intern. Vacuum Congress and 5th Intern. Conf on Solid Surfaces, Invited Speakers Volume (ASEVA, Madrid 1983) ed. by J.L. de Segovia, p. 3
5. D.W. Bassett: In Surface Mobilities on Solid Materials, ed. by V.T. Binh (Plenum, New York 1983) p. 63, 83
6 A recent review has been given by N. Ernst, G. Ehrlich: In Microscopy Methods in Metals, ed. by U. Gonser, Topics Current Phys. 40, (Springer, Berlin, Heidelberg 1986) p. 75, 115
7. H.-W. Fink, G. Ehrlich: Surface Sci. 143, 125 (1984)
8. G. Ayrault, G. Ehrlich: J. Chem. Phys. 60, 281 (1974)
9. D.W. Bassett, P.R. Webber: Surface Sci. 70, 520 (1978)

10 J D Wrigley, G. Ehrlich: Phys Rev Lett. 44, 661 (1980)
11 H. Wagner: In Springer Tracts Mod. Phys. 85, 151 (Springer, Berlin, Heidelberg 1979)
12. G. Ehrlich, K Stolt: In Growth and Properties of Metal Clusters, ed by J. Bourdon (Elsevier, Amsterdam 1980) p. 1
13. H -W Fink, G Ehrlich: Surface Sci. 173, 128 (1986)
14. See, for example, A A. Chernov Modern Crystallography III Crystal Growth, (Springer, Berlin, Heidelberg 1984)
15. H -W Fink, G. Ehrlich. Phys. Rev. Lett 52, 1532 (1984)
16 H.-W Fink, G. Ehrlich: J Chem Phys 81, 4652 (1984)
17 J.P Bourdin, M C Desjonqueres, D Spanjaard, J Friedel Surface Sci. 157, L345 (1985)
18 H. Dreysse, D Tomanek, K H. Bennemann Surface Sci. 173, 538 (1986)
19. H -W Fink, G. Ehrlich Surface Sci. 150, 419 (1985)

Topography Modification and Microscopic Motion on Metal Surfaces

*R.J. Behm**

Fritz-Haber-Institut der Max-Planck-Gesellschaft,
Faradayweg 4–6, D-1000 Berlin-Dahlem (33), Germany

1. Introduction

Overall mass transport, as described by diffusion laws, is - in a microscopic picture - based on the more or less independent motion of single particles such as atoms, molecules or larger agglomerates /1/. In solid, ordered phases or at their surfaces this mobility can also lead to structural modifications, which are not necessarily connected with (long-range) mass transport, for example in ordering processes. The role and mechanism of particle migration in a number of surface processes on single-crystalline metal surfaces will be the subject of this article. It is demonstrated that, from examination of the structural modifications of these surfaces on an atomic scale, it is possible to draw conclusions on the extent and mechanistic details of particle transport during these processes. This work is primarily based on structural data which resulted from Scanning Tunneling Microscopy (STM) investigations of surfaces performed in an UHV environment /2/. In situ Auger Electron Spectroscopy (AES) and Low Energy Electron Diffraction (LEED) served to control surface cleanliness and to establish a link with results of earlier studies.

Reactive surface processes often involve structural rearrangements of the surface layer itself. In many of those reactions the mobility of the metal atoms in the topmost layer actually represents the rate-limiting step. Consequently this contribution will focuss on the mobility of surface metal atoms, which can be treated, however, in the standard framework used to describe the mobility of adsorbed particles. The topmost metal layer is taken as an adlayer adsorbed on the metal substrate.

The displacement of an adsorbed particle on a (single crystalline) surface can occur in different ways. In one case the particle continuously remains on the surface and moves by hopping from one adsorption-site to another. Here the mobility is determined by the effective hopping rate resulting from an activation barrier E_h^* and a frequency factor ν_h (fig. 1a). In a second case, which is not effective in the processes discussed here, particles are displaced by continuous evaporation and readsorption, i.e. by transport via the adjacent vapor phase. Here the evaporation rate would be governed by the activation barrier for desorption E^*_d plus a frequency factor ν_d.

The hopping process of a single adparticle and its rates are sensitively affected by interactions (fig. 1b) between neighboring adparticles and

*pres. addr.: Institut für Kristallographie und Mineralogie, Universitat Munchen, Theresienstr. 41, D-8000 Munchen 2, Fed. Rep. Germany

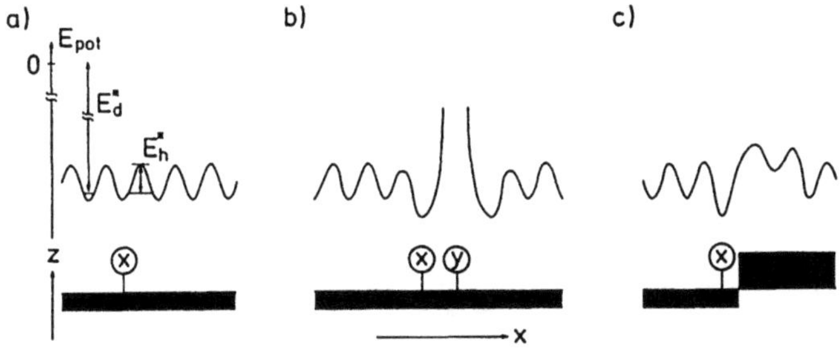

Fig. 1 Potential energy (top) and topographical scheme (bottom) of an adparticle on a one-dimensional, periodic surface. E_d^*: activation energy for desorption; E_h: activation energy for migration; a) single adparticle on an ideal surface; b) adparticle in the presence of a neighboring adparticle; c) adparticle in the presence of a surface defect (step)

especially by surface inhomogeneities like steps, point defects or chemically different atoms in or on the surface (fig. 1c). This picture gives an indication of the influence the surface structure can exert on the mobility of adparticles. In the case of metallic adatoms on a metal substrate the cohesive interactions with all adlayer metal atoms in the neighborhood can be summarized as the two-dimensional (2D) condensation energy. In contrast to other adsorbates, where typical interaction energies are around 1 kcal/mole, these interactions can be expected to be relatively strong. Surface mobility of adatoms which are chemically identical to the substrate, like in the examples presented here, lead to self-diffusion.

2. Topography of Clean Metal Surfaces

The defect structure of a clean and well annealed metal surface provides a first insight into the potentials felt by a migrating metal atom: Point defects such as missing atoms in the topmost layer or single adatoms have never been detected in STM studies, and also smaller islands of adsorbed metal atoms become more and more rare with increasing annealing temperature /2-5/. The empty sites of single missing atoms apparently represent very effective "traps" for migrating adatoms, corresponding to energetically very deep states. In analogy, single adatoms must be very unstable; they will be trapped during their random walk over the substrate surface whenever they hit a trap - e.g. missing atoms, steps, etc. The observed surface structure thus indicates that for the migration of metal atoms on a metal substrate the mobility is dependent on the affinity of trap sites such as steps and kinks rather than on the hopping rate of adatoms and its activation barrier /6/.

3. Island dissolution/condensation

The role of steps and kinks as active sites for two-dimensional evaporation and condensation is evident in the processes of island dissolution and island growth.

The dissolution of monolayer Pt islands on a Pt(100) surface was in situ monitored by STM at 700 K /7/. The islands had resulted from the adsorbate-induced structural transformation of the hexagonal reconstructed Pt(100) surface into the bulk-like (1x1) phase /8/. More details of this process will be discussed below. In this experiment ethene adsorption was used to

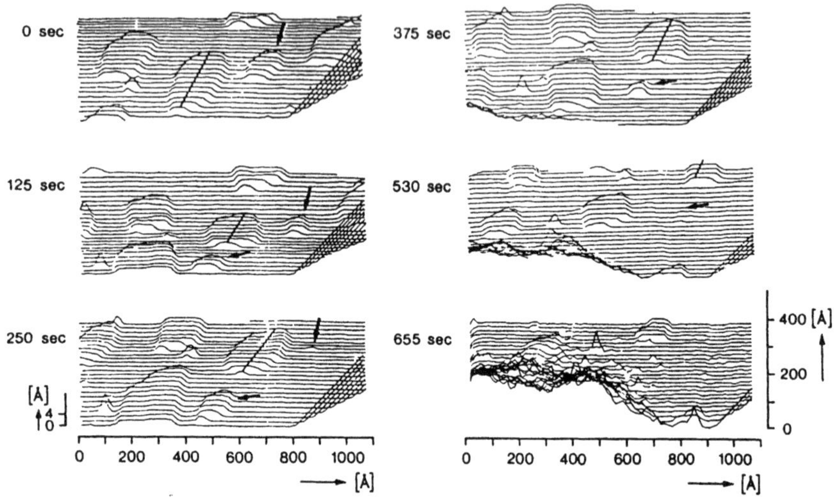

Fig. 2 Dissolution of monolayer Pt islands on a Pt(100) substrate at 700 K, recorded in a sequence of STM images (timescale marked left)

impel the transformation; the Pt surface was consequently covered by an adlayer of decomposed ethene /9/. The series of images in fig. 2 was taken one after the other with the time-scale indicated at each pattern. The first pattern displays a number of neighboring Pt islands on an otherwise smooth background. The second pattern, taken 125 sec later, shows almost the same surface area but slightly displaced and distorted due to thermal drift, which in the first pattern moves opposite and in the second parallel to the direction of the slow scan. Also the ensuing patterns in this sequence cover the same area of the surface except for slight shifts related to thermal drift. This latter effect is responsible for the irregular features entering the scanned area in patterns 4-6. For better orientation one identical island is marked by a diagonal line throughout this series. On closer inspection one sees that there are also small changes in topography in addition to distortion effects. In comparing patterns 1 and 2 the island marked by an arrow in the first pattern is seen to have decreased in size. In the following pattern taken after 250 sec this island has almost completely disappeared. A similar process is observed for another island marked by an arrow from the second pattern on. The island decreases in size between patterns 3 and 4 and has finally disappeared after 530 sec. At the same time the large island left of the marked one loses material at its left side, leading to an increased distance between island perimeter and the small island at the lower-left corner of the patterns. These patterns indicate that for a Pt(100) surface covered by ethene fragments at 700 K, the evaporation rate from monolayer Pt islands is such that island dissolution can occur in the timescale of the experiment, over several hundreds of seconds.

These images allowed direct observation of the change in island perimeter and size and their final dissolution. They did not reveal where the adatoms have gone. One would expect them to condense at other steps in such a way that overall large islands grow at the expense of smaller ones. Though not directly monitored, this could be concluded from a sequence of annealing experiments on similar surfaces with increasingly higher an-

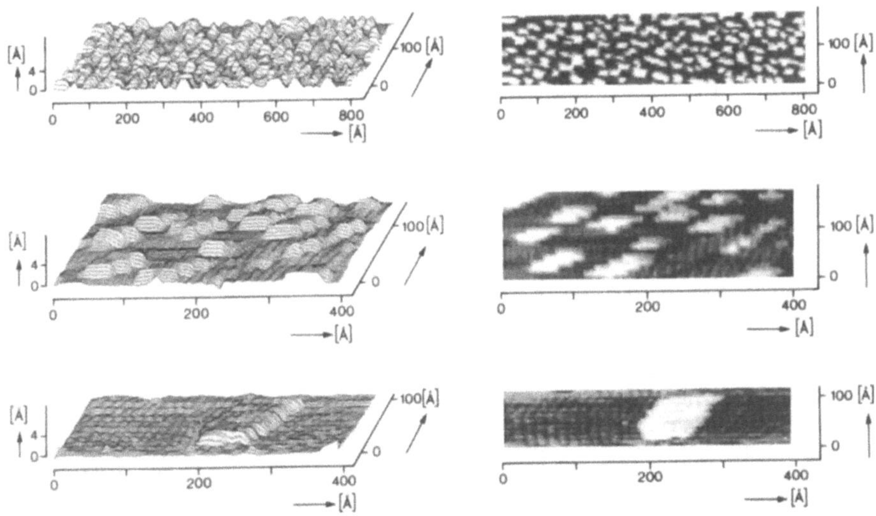

Fig. 3 Pt islands after adsorption of CO/NO at 370 K (a), typical image) and following subsequent annealing to 445 K (b), typical image) 495 K (c), maximum island size); STM line patterns (left) and top-view in grey scale representation (right)

nealing temperature. In contrast to the first series the island covered Pt(100) surface had been produced by adsorption of CO or NO /10/. Both of these adsorbates desorb at T ≥ 500 K and leave a clean surface; NO dissociates prior to desorption. The STM images showed that the size of the islands increased with annealing temperature, whereas their number decreased drastically /8/. The adatoms set free from the perimeter of melting islands thus have indeed condensed to larger islands. This behaviour is displayed in a few typical images of that sequence, selected from a large number of experiments (fig. 3). The temperatures at which the clean surface was annealed to in this sequence, are significantly lower than the temperatures at which the ethene covered surface was kept throughout the entire experiment. This apparent contradiction is discussed in more detail in the context of chemical effects in section 4 of this contribution.

As demonstrated by the above results, the cohesive interactions can significantly stabilize adsorption of metallic adatoms at step edges. This phenomenon could also serve to explain the topographic effects observed in the structural transformation on Pt(100) already mentioned. Following the adsorbate-induced structural transformation 'hex' reconstructed → (1x1) of a formerly flat surface this was now covered by small islands /8/. Their height of 2.0 Å, identical to the interlayer spacing of Pt(100), leads one to ascribe them to monolayer Pt islands. Their width varied, depending on the transition temperature, from ~ 25 Å at 370 K to 70 Å - 100 Å at 470 K /8/. This observation must be seen in the context of other experimental observations. The structural transformation itself has been reported from Rutherford Backscattering (RBS) experiments to occur at least down to temperatures around 150 K /11/. In contrast the resulting LEED pattern indicates a poorly ordered surface up to 400 K /10/. It was shown that these features can be understood in terms of a two-step mechanism /8, 12/. In the first step the actual structural transition takes place. The concomitant density reduction of ~ 20 % in the topmost layer /13/ leads to

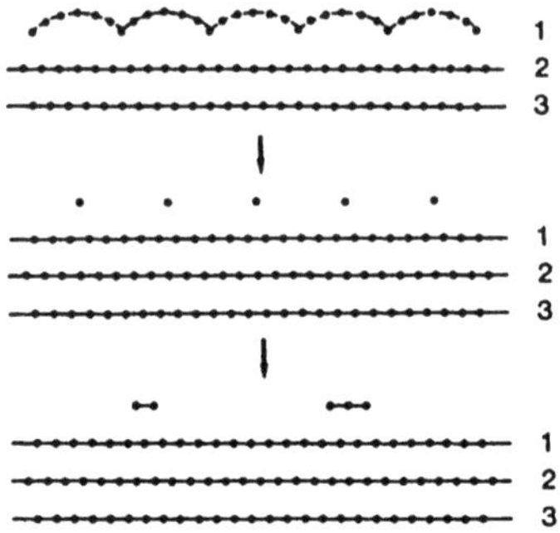

Fig. 4 Schematics of the two-step mechanism proposed for the adsorbate induced structural transformation 'hex' reconstructed → (1x1) on Pt(100). Lines indicate layers of Pt atoms (full line: bulk structure, dotted line: reconstructed layer), circles represent single Pt atoms

an excess of surface atoms, and these extra Pt atoms are locally expelled into the next layer (fig. 4). In the next step those atoms can migrate over the surface and either become trapped at steps etc. or condense into little islands, as observed by STM. The size of the newly created islands determines the long-range order as probed by LEED, while the (1x1) short-range order is observed through RBS. The low temperature of 150 K sufficient for the actual transformation to proceed indicates this step to have a rather low activation barrier. The temperature dependence of the island size, in contrast, points to a measurable barrier for the step determining the island size.

More quantitative results about this second step came from a recent Monte Carlo simulation /14/. Therein the process of island formation was modelled by adding one adatom after another in a random mode, while in between the already adsorbed atoms were allowed to perform a fixed number of attempted migration steps (10-4000). Attractive pair-wise interactions ε between adatoms on nearest neighbor (n.n.) sites represented a first order approximation for the 2D condensation energy.

The correlation between average island size and coverage for different values of the reduced n.n. attraction ε/kT is depicted in fig. 5. Interestingly enough, this approximation not only agreed reasonably well in the absolute island size for interaction energies of the same magnitude as reported from other experiments /15/; it also pointed out that the island size was reduced in different ways by too high or too low interaction energies, leaving only a limited range of "favorable" interaction energies in between. At too high values of ε/kT, those particles, which were once condensed into small islands have almost no chance of re-evaporation. As a consequence it was not possible for the larger islands to grow at the expense of the small ones. The surface was covered with many rather small islands ("kinetic limitation"). If, on the other hand, the attractive forces were too small the islands became thermodynamically unstable and their average size dropped ("thermodynamic limitation"). In this case also a 2D vapor phase existed which consisted of many single atoms and which was not present under kinetic limitations. For the reaction modelled here the system is - at least at lower temperatures - within the range of kinetic limitations: going to higher temperatures leads to an increase in island size, as is expected.

Although the numerical results of that simulation should not be overstressed, the model nevertheless provides valuable insight into the overall mechanism of the above structural transformation. It demonstrates that the cohesive energies between adatoms are sufficient to explain the size of the

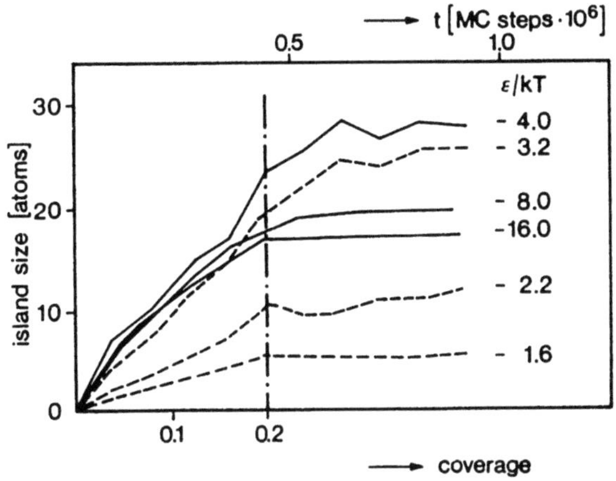

Fig. 5 Development of the average island size for Pt/Pt(100) in a kinetic Monte Carlo simulation (4000 migration steps per new particle) as a function of coverage. Region at $\theta > 0.20$ indicates the relaxation at this coverage in another $2 \cdot 10^6$ migration steps. Pairwise reduced interaction ε/kT energy as indicated

observed islands and its temperature dependence and thus does not contradict the proposed mechanism. The hopping rate is, except for the very low rates around 10 migration steps per new particle, of negligible influence. This simulation again emphasizes the role of steps and kinks as active sites in surface mobility of metallic surfaces in comparison to the hopping rate of metal adatoms. Finally a specific and significant effect of kinetic limitations was introduced and discussed here.

4. Chemical stabilization/destabilization

The role of chemical inhomogeneities, such as foreign atoms in/on the surface layer, upon the mobility of adatoms has not been systematically investigated yet. Nevertheless a number of STM studies in our laboratory on different surfaces so far have indicated that incorporated or strongly adsorbed atoms like oxygen or carbon atoms tend to stabilize steps on metal surfaces, i.e. they prevent the evaporation from terrace edges. It was observed that in those cases the terrace size even of well annealed surfaces was much smaller than the ones found on clean surfaces.

The variation in island dissolution rate and thus in the step stability caused by the presence of an adsorbate layer was likewise visible when comparing two of the experiments described above. In the case of the 'in situ' observation of island dissolution on a Pt(100) surface covered by decomposed ethene, this process was relatively slow at 700 K. On the clean Pt(100) surface, in contrast, it proceeded at a much faster rate. One annealing cycle to 500 K was sufficient to remove most of the islands. Apparently the adsorbate layer stabilizes the surface. In the terminology used above, that means it reduces the evaporation rate from step edges.

Another kind of experiment probed the stability of very small metal clusters containing foreign atoms. A smooth Pt(100) surface was exposed to a small amount of ethene and then heated to 1100 K /2,4/. STM images taken after cool-down revealed that the surface had become covered by small

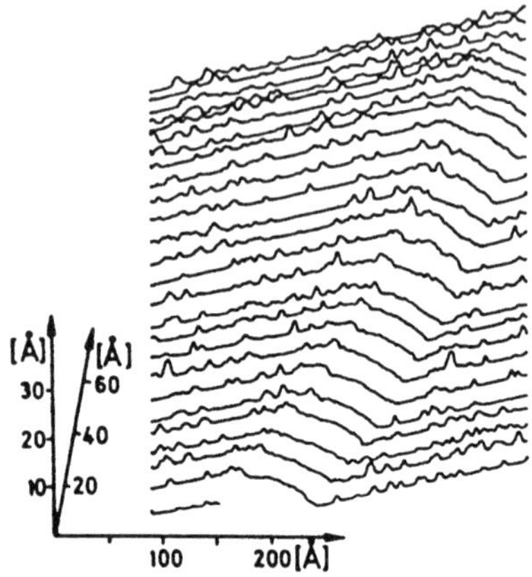

Fig. 6 STM image of a Pt(100) surface with small amounts of adsorbed carbon after annealing to 1100 K and subsequent cool-down

islands which from their vertical extension (multiple of Pt(100) interlayer distances) could be identified as small 2D platinum clusters. This experiment was interpreted in such a way that the carbon atoms are incorporated into small Pt clusters and stabilize those clusters. The clusters must have been formed by Pt atoms which were mobile at the high annealing temperatures and migrated freely over the surface before being condensed into these stabilized clusters. This behaviour is comparable to bulk effects where incorporated atoms are widely known to stabilize the lattice, i.e. to act as "pinning centers". It is important to note that these islands are only formed after high temperature annealing of the carbon-precovered surface, which yields a sufficiently high concentration of free Pt adatoms. The role of the adsorbate in this case can be described as that of providing nucleation centers for the island formation.

5. Domain Mobility

The foregoing examples illustrated how the mobility of single metallic adatoms was affected by structural and chemical inhomogeneities and how the surface topography was shaped by processes involving adatom mobility. The mobility of single adatoms is especially important for the long-range transport, the (re-)formation of ordered layers in contrast depends more on the mobility within an adlayer. Especially in rather close packed adlayers this may be markedly different from that of single adatom mobility. The matrix of neighbors makes changes in the adsorption site much less favorable and thus leads to a localization of the adatoms.

Such an effect is directly deduced from STM observations of the thermally activated transformation of the metastable (1x1) into the thermodynamically stable 'hexagonal' reconstructed phase of the clean Pt(100) surface. Due to the symmetry of the fcc(100) substrate, four different rotational orientations of the topmost reconstructed layer are possible /13/. In the course of the structural transformation the new phase develops in many small domains with different orientations /16/. While it had been shown that terraces of a well annealed surface are made up of only one domain , different domains were shown to exist when the surface was kept at lower temperatures (<600 K) /12/. These domain boundaries still existed after annealing to temperatures where the islands described above have long been dissolved. The difference in annealing temperatures - < 600 K and ~ 1100 K /16/ - implies considerably different rates for these two processes, island dissolution (2D evaporation) and reorientation of close-packed domains of adatoms, in accord with the effect of localization by surrounding adatoms described above.

6. Reaction Induced Topography Modification

This section is devoted to adlayer mobility in the course of surface reactions, which is closely related to the effects of chemical stabilization as described in section 4. There the thermally induced mobility of the surface layer was tested while its chemical composition was kept constant. In the present case it is just the opposite: the temperature of the surface is kept constant while the chemical composition is changed continuously, e.g. by exposure to reactive gases. Results of a recent STM study on the interaction of oxygen with a Ni(110) surface /5/ will serve as an example for the reaction-induced mobility of the adlayer. In view of the foregoing results, interest will again be focussed on the adsorbate-induced variation in step stability, i.e. on the evaporation rate of metallic adatoms in the presence of the adsorbate.

In extended series of STM images it was found that at $T \leq 470$ K the clean Ni(110) surface was completely stable. It consisted predominantly of rather large, flat terraces separated by mostly monoatomic steps. No significant changes could be detected in the terrace/step structure. Exposure to small amounts of oxygen caused the surface to reconstruct into the well known (2x1) and (3x1) structures /17/, which was verified in situ by LEED. The STM observations showed that in this coverage regime and at temperatures $340 \text{ K} < T < 470$ K the step/terrace topography of the surface was maintained throughout the adsorption process. Besides structural modifications due to the formation of the (2x1) and (3x1) structures, the terraces did not change in their general topographical appearance. No islands of extra atoms nor any other new kind of structural elements were detected that could indicate the accommodation or loss of metal atoms. Therefrom the authors concluded that these reconstructions cannot result from structures exhibiting a mass deficit in the topmost layer as compared to the clean surface. By this argument the often favored 'missing row' model for the (2x1) reconstruction /18/ was strongly questioned.

Higher exposures, which were again in situ calibrated by LEED using the unique pattern of the (9x4) suboxide structure /19/, led to significant modifications of the surface topography. This effect is displayed in a set of STM images taken during oxygen adsorption at 420 K (figs. 7a-d). The clean surface in fig. 7a is characterized by an extended terrace and a monoatomic step. This picture did not change throughout the coverage range of the (2x1) and (3x1) structures over a series of 7 images. At the onset of suboxide formation the formerly flat terrace suddenly grew a depression in its upper right part, as if part of the topmost layer were eroded away. The characteristics of surface erosion became even more pronounced at higher exposures (figs. 7c,d). Increasing amounts of the upper terrace were eroded away. In order to eliminate effects of thermal drift the right section of fig. 7 shows just part of these images such that always the same area is displayed. For better identification the areas on the level of the lower terrace are marked black. Clearly these eroded areas grow with increasing coverage starting either from the initial nucleation center in fig. 7b or from the step in fig. 7a: The structural implications of this process were discussed in detail in ref. 5. The important point in the context of this section is the fact that in the coverage regime of suboxide formation the adsorbed/incorporated oxygen atoms lead to a destabilization

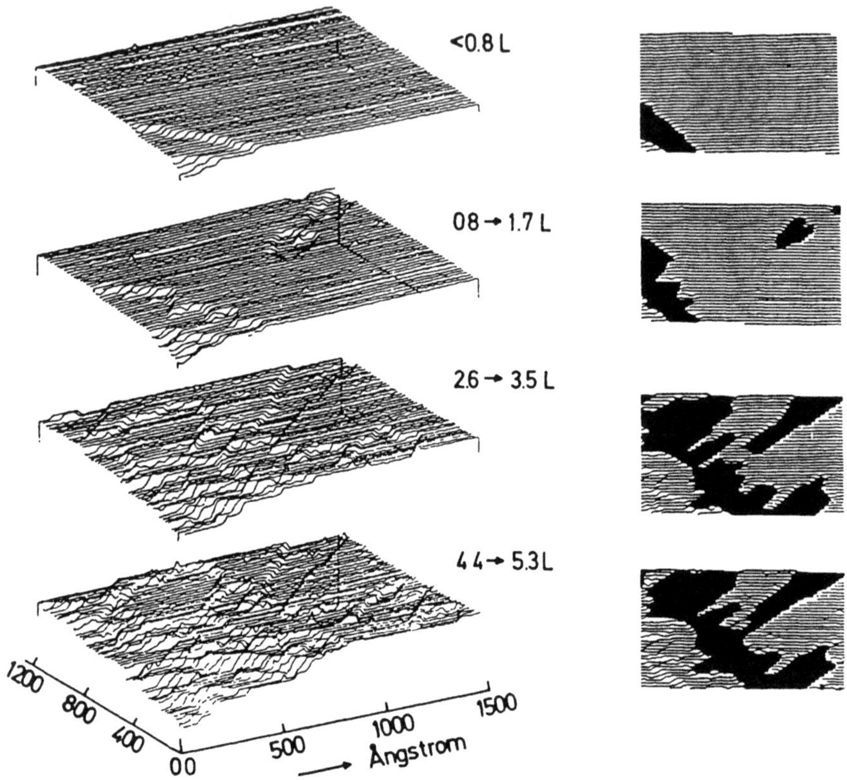

Fig. 7 Sequence of STM images (left) taken during oxygen adsorption on Ni(110) at 420 K, ranging from the clean surface to suboxide formation; areas on the lower terrace level are marked black in the (drift-corrected) topview of a constant section of these images (right); oxygen exposure is indicated in the center

of the steps such that Ni adatoms can leave the upper terraces (2D evaporation). This is in striking contrast to the behavior of the clean surface as well as to that of the oxygen covered surface in the (2x1) and (3x1) reconstructed phases. It implies that in the suboxide the oxygen atoms significantly weaken the cohesive forces between metal atoms in the topmost layer. This effect works in the opposite direction as in the system C/Pt(100) described above where the C atoms stabilize metal clusters. One should not forget, however, that this increase in step mobility was observed only during the extensive rearrangement of the lattice related to the suboxide formation, which is presumably connected with a break of part of the metallic bonds in the topmost layer /19/.

Summary

Based on STM images it was shown that surface mobility as an integral part of many processes on (single-crystalline) metal surfaces often results in or depends on characteristic structural elements on that surface. The cohesive forces between surface layer metal atoms and their variation by the presence of chemically different adparticles were found to play a dominant role as compared to the single-particle hopping process. These forces are effective at step edges, e.g. in island nucleation/condensation, in island dissolution or terrace modification, and disclose step/kink sites to be most important for this type of surface mobility. It was illustrated how microscopic, atomic-scale observations of the structural modification of a surface in a given process can provide insight into the details of particle transport and the underlying mechanism. These details can be used for a quantitative evaluation via Monte Carlo simulations, as shown for the structural transformation of a Pt surface. This procedure also permits a test of the correctness of the proposed model. The knowledge of microscopic details can thus form the basis for a qualitative and even quantitative understanding of a surface process.

ACKNOWLEDGEMENTS
The good collaboration with E. Ritter, J. Wintterlin (experiments) and A. Reynolds, D. Kaletta (Monte Carlo simulations) is gratefully acknowledged. Financial support for this work came from the Deutsche Forschungsgemeinschaft (SFB 128).

REFERENCES

1. H. J. Kreuzer: this volume
2. W. Hösler, R. J. Behm, E. Ritter: IBM J. Res. Develop. 30, 403 (1986)
3. G. Binnig, H. Rohrer, C. Gerber, E. Weibel: Surf. Sci. 131, L379 (1983)
4. R. J. Behm, W. Hosler: In Chemistry and Physics of Solid Surfaces VI, ed. by R. Vanselow and R. Howe (Springer, Berlin, 1986)
5. E. Ritter, R. J. Behm: In Proc. ICSOS II (Amsterdam, Netherlands, 1987)
6. J. M. Blakely and M. Eizenberg: In The Chemical Physics of Solid Surfaces and Heterogeneous Catalyis I, ed. by D. A. King and D. P. Woodruff (Elsevier, Amsterdam, 1981);
   G. Ehrlich: In Chemistry and Physics of Solid Surfaces V, ed. by R. Vanselow and R. Howe (Springer, Berlin, 1984)
7. E. Ritter: Ph.D. Dissertation, Univ. München (1987)
8. E. Ritter, R. J. Behm, G. Potschke, J. Wintterlin: Surf. Sci. 181, 403 (1987)
9. K. Baron, D. W. Blakely, G. A. Somorjai: Surf. Sci. 41, 157 (1974)
10. R. J. Behm, P. A. Thiel, P. R. Norton, G. Ertl: J. Chem. Phys. 78, 7437 (1983), ibid. 7448
11. P. R. Norton, J. A. Davies, D. K.Creber, C. W. Sitter, T. E. Jackman: Surf. Sci. 108, 205 (1981)
12. R. J. Behm, E. Ritter, W. Hosler: in preparation
13. M. A. van Hove, R. J. Koestner, P. C. Stair, J. P. Biberian, L. L. Kesmodel, I. Bartos, G. Somorjai: Surf. Sci. 103, 189 (1981)
14. A. E. Reynolds, D. Kaletta, R. J. Behm, G. Ertl: to be published
15. L. Roelofs: In The Physics and Chemistry of Solid Surfaces IV, ed. by R. Vanselow and R. Howe (Springer, Berlin, 1982)
16. R. J. Behm, W. Hosler, E. Ritter, G. Binnig: Phys. Rev. Lett. 56, 228 (1986)
17. C. R. Brundle, J. Broughton: In The Chemical Physics of Solid Surfaces and Heterogeneous Catalysis IIIB, ed. by D. A. King and D. P. Woodruff (Elsevier, Amsterdam, 1988)
18. H. Niehus, G. Comsa: Surf. Sci. 151, L171 (1985)
19. J. W. May, L. H. Germer: Surf. Sci. 11, 443 (1968)

Surface Self-Diffusion, Capillarity, and Surface Steps

J.M. Blakely and C.C. Umbach

Dept. of Materials Science and Engineering,
Cornell University, Ithaca, NY 14853, USA

A brief description of surface self-diffusion at crystal surfaces is presented. Surface fluxes are defined as excess quantities and possible mechanisms of diffusion are discussed involving surface point defects. The connection between macroscopic capillarity treatments of surface shape development by diffusion and the dynamics of evolution of surface atomic step arrays is discussed.

1. Surface Flux and Diffusivity

Surface diffusion coefficients can be defined through flux equations in a manner similar to that used for diffusivities in bulk phases. The diffusion coefficient is defined to be the proportionality factor between the flux and concentration gradient as expressed by Fick's first law. A convenient formulation can be made in terms of a surface *excess* flux. Consider the situation depicted in Fig. 1(a) of a semi-infinite slab of material with a gradient in the chemical potential, μ, of the species of interest in one direction only, parallel to the surface. The surface excess flux J_s of that component (atoms/cm/sec) can then be defined as

$$J_s = \int_0^\infty [J(y) - J_{bulk}]dy \tag{1}$$

where J_{bulk} is the flux in the interior of the solid well away from the surface and J(y) its value at the depth y below the surface, Fig. 1(b). J_s defined in this way is an integral quantity. The depth dependent flux J(y) can in turn be written in terms of a depth dependent diffusivity D(y) as

$$J(y) = -\frac{[n(y)D(y)]}{kT}\nabla\mu \tag{2}$$

where n(y) is the concentration of diffusing atoms at depth y so that

$$J_s = -\left(\int_0^\infty [n(y)D(y) - n_{bulk}D_{bulk}]dy\right)\frac{\nabla\mu}{kT} \tag{3}$$

It is customary in the literature to write this in terms of an *effective* number n_s of diffusing atoms per unit area of surface with diffusivity D_s, i.e.

$$J_s = -n_s D_s \frac{\nabla\mu}{kT} \tag{4}$$

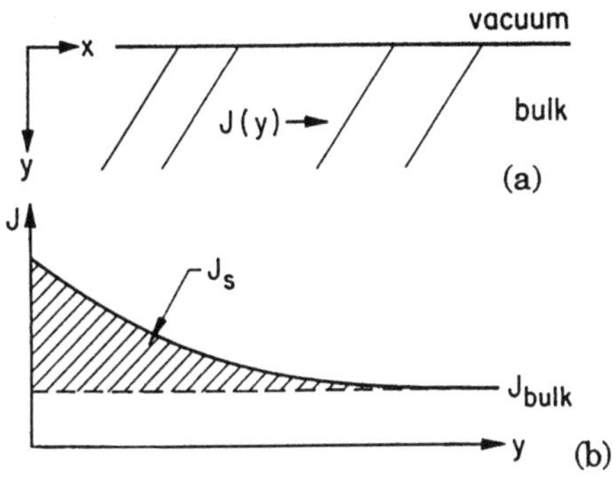

Fig. 1. Diagram illustrating the definition of the surface diffusive flux as a surface excess quantity. In (a) the gradient in chemical potential is taken to be parallel to the surface only. The shaded area in (b) depicts the surface excess flux.

For the case of diffusion within an adsorbed layer only, the bulk flux is of course zero and n_s is the actual number of adsorbed atoms per unit area. For mass transport processes involving self-diffusion at crystal surfaces the contributions of many atomic layers may be important; the diffusivity D(y) would be expected to reach its bulk value after only a few monolayers for metals and covalent crystals but for ionic crystals the transition region may be much greater (10^2 to 10^3 atomic layers) /1/.

2. Mechanisms of Surface Diffusion

2.1 Monolayer Models

The average jump frequency of atoms in the top monolayer of a crystal is expected to be quite different from that of bulk atoms. Figure 2 depicts a surface vicinal to some low index plane and shows atoms in a variety of crystal sites--terrace, ledge, kink, and adatom. For a given orientation we require some minimum density of terrace, ledge, and kink sites but at elevated temperatures excited states of the surface may involve an increased kink density and the appearance of adatoms and terrace vacancies. As long as the densities of these extra defects are small they may be regarded as independent with densities described by simple Boltzmann factors. Cooperative 'roughening' /2,3/ is the phenomenon expected at elevated temperatures when the interaction among the spontaneously created defects is important.

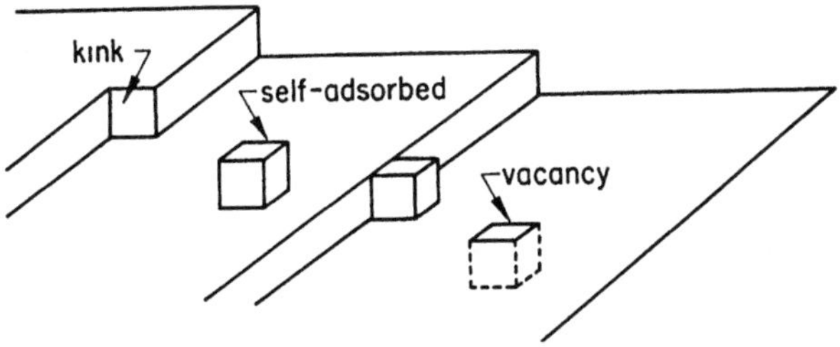

Fig. 2. Some of the elementary defects expected on a vicinal surface

The self-diffusion coefficient of the surface monolayer atoms in the low defect density limit can be written as

$$D_s = \sum_i f_i \left(\frac{n_i}{n_s} \right) D_i \qquad (5)$$

where n_i is the number of defects of type i and f_i is the correlation factor that accounts for the interdependence of successive jumps. In writing this equation we assume that surface atoms only diffuse via excited defect states and that processes such as place exchange are unimportant; the latter would contribute to tracer diffusivities but not to mass transport. The point defect diffusivity can in turn be expressed as

$$D_i = \left(\frac{1}{4} \right) \Gamma_i a_i^2 \qquad (6)$$

where Γ_i is its average jump frequency and a_i is its average jump length; for surfaces with less than threefold rotational symmetry about their normals the tensor character of D_i would appear in the directional dependence of Γ_i and a_i. The measurement of the parameters determining D_i is discussed by FINK in the paper in this volume. For the case of of an individual adatom diffusing across terrace regions the variations of crystal free energy with position would be periodic with a barrier of height ΔG_m between successive equivalent sites (Fig. 3). For $\Delta G_m >> kT$ the adatoms would be localized in self-adsorption sites for most of the time and would occasionally undergo a hopping motion. For $\Delta G_m < kT$ the adsorbed atoms would behave like a two-dimensional ideal gas. The available evidence on barrier heights from field-ion microscopy (FIM) experiments on refractory metals /4,5/ suggests that the localized adsorption model will generally be more appropriate but the possibility of delocalization at high temperatures cannot be ruled out, especially on anisotropic materials (such as graphite) where most of the binding energy can be accounted for by in-plane bonds. In the localized adsorption limit,

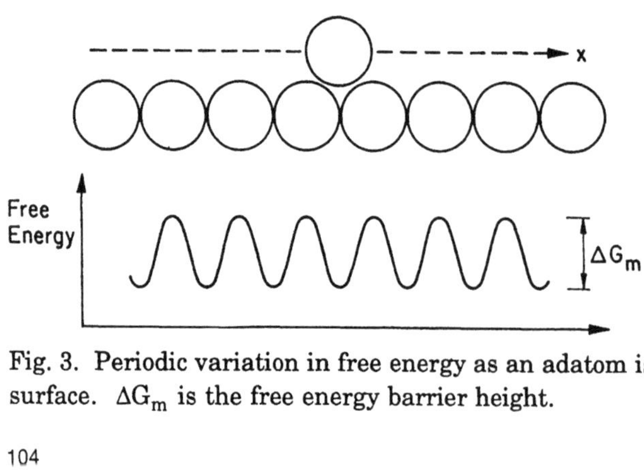

Fig. 3. Periodic variation in free energy as an adatom is translated over the surface. ΔG_m is the free energy barrier height.

$$\Gamma \approx \upsilon \exp\left[-\frac{\Delta G_m}{kT}\right] \qquad (7)$$

in the usual 'quasi-equilibrium' diffusion model, with υ a characteristic vibrational frequency of the defect.

In the low adatom density limit,

$$\frac{n_i}{n_s} = \frac{1}{\left\{1+\exp\left[\frac{(G_{ad}-\mu)}{kT}\right]\right\}} \approx \exp\left[-\frac{G_{ad}-\mu}{kT}\right] = \exp^{-\frac{\Delta G_f}{kT}} \qquad (8)$$

where μ is the atom chemical potential, G_{ad} the free energy when an adsorption site is occupied, and ΔG_f is the free energy of formation of the adatom.

Combining (5) through (8) gives the surface diffusivity for the adatom mechanism as

$$D_s = \frac{1}{4}\upsilon f a^2 \exp\left[\frac{-(\Delta G_f+\Delta G_m)}{kT}\right] \qquad (9)$$

and hence D_s has the form

$$D_s = D_0 \exp^{-\frac{Q}{kT}} \qquad (10)$$

with Q the overall activation energy.

As noted earlier this simple description would only be appropriate in the region of low defect densities. When interactions are important the defect densities would increase more rapidly with temperature than described by (8) and non-Arrhenius behavior would be obtained for D_s. Surface roughening /6/ and surface melting /7/ would both be expected to cause this type of deviation.

2.2 Multilayer Diffusion

In ionic materials with very low densities of mobile charge the effects of free surfaces may extend to many layers into the crystal. The theory of surface ionic space charge regions is quite well developed /8,9,10/ although the amount of experimental data on such layers is rather meager. The binding energies associated with surface cations and anions will generally differ from those for bulk sites and any imbalance in the densities of the two ions will lead to a net surface charge balanced by a subsurface defect space charge region. In such materials, therefore, the effective surface diffusivity, as defined by (4) will have contributions from both the surface monolayer and the subsurface region.

Materials other than ionic solids may also have atomic defect concentrations and mobilities that vary over several atomic layers beneath the surface but with metals and semiconductors the effects are likely to be of much shorter range.

3. Mass Transport by Surface Diffusion

3.1 Macroscopic Description

The macroscopic theory of the evolution of surface shapes with time is based on the Gibbs-Thomson relationship /11/

$$\Delta\mu = K\gamma\Omega \tag{11}$$

for the difference in surface potential of atoms at a portion of the surface with local curvature K relative to a flat region with K=0. γ is the surface tension and Ω is the atomic volume. This equation assumes the surface tension γ to be orientation independent; this will rarely be the case but the approximation may be justified when γ is a weak function of orientation or the range of surface orientations involved does not include ones corresponding to sharp minima of γ. The effect of non-spherical Wulff plots on the Gibbs-Thomson relation is discussed by HERRING /11/ in a now classic paper on the general topic of mass transport promoted by capillarity. In the isotropic case there will be a net flux of atoms along the surface from regions of high curvature to regions of low curvature. The surface flux can be written formally as

$$J_s = -\frac{n_s D_s \gamma\Omega}{kT}\nabla K \tag{12}$$

(There will also be contributions to observed shape changes from other modes of mass transport such as volume diffusion and evaporation-condensation. These contributions must be accounted for in the interpretation of data but can be minimized by using structures of very small dimensions.) MULLINS and associates /12/ made an important contribution to surface mass transport theory by developing equations for particular surface shapes and hence provided a link to experimental work. For the very special case of a surface of sinusoidal shape (Fig. 4), where

$$y = A\sin(2\pi x/\lambda) \tag{13}$$

the development of (12) leads to a time-dependence of

$$y = y_0 \exp(-\alpha t) \quad \text{where} \tag{14}$$

$$\alpha = \left(\frac{n_s D_s \gamma\Omega}{kT}\right)\left(\frac{2\pi}{\lambda}\right)^4 \tag{15}$$

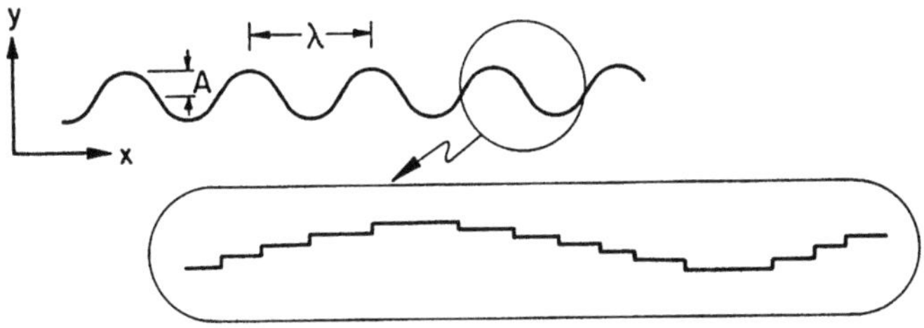

Fig. 4. Sinusoidal profile of a crystal surface. Inset corresponds to monoatomic step distribution when the average surface is a low index plane. The maximum step density corresponds to the inflection point of the sine wave.

The surface remains sinusoidal with the amplitude decaying exponentially with time. Surface sine waves were first produced and measured by BLAKELY and MYKURA /13/ and were shown to be well described by the Mullins analysis. The sine wave or grating decay method has subsequently been developed and used by several researchers /14,15/ on a variety of materials, including metals /15/, Si /16/, and a few ionic solids /17,18,19/.

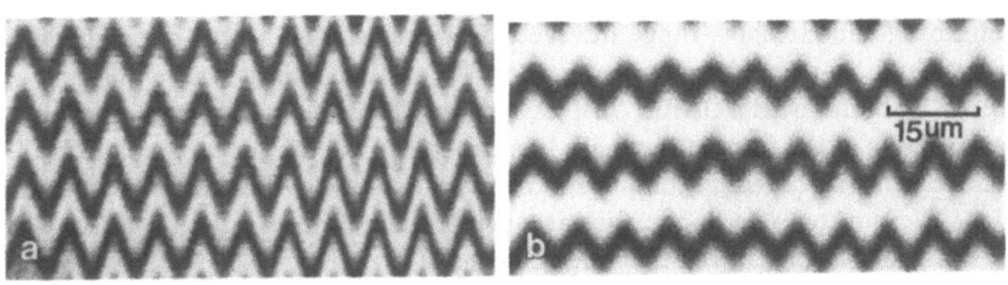

Fig. 5. Interference micrographs from a sinusoidal Al_2O_3 surface with mean orientation $(10\bar{1}0)$ /19/. Annealing for 168 hr. at 1650 C between (a) and (b) has reduced the amplitude from approximately 70 to 40 nm. The grating spacing is 7.5 μm. (The indicated lateral scale is the same for both micrographs.)

Figure 5 shows an example of the decay of a sinusoidal Al_2O_3 surface; a variety of different methods have been used for monitoring the changes in such surfaces including interference microscopy /13/, optical diffraction /20,21/, profilometry /16/ and low energy electron scattering /22/.

3.2 Microscopic Description

The case of the development of the sinusoidal surface is a convenient example in terms of which to discuss the connection between the macroscopic capillarity treatment and elementary atomic transport processes. For crystal surfaces the shape can in principle be specified by giving the arrangement or spatial distribution of atomic steps. For the one-dimensionally sinusoidal surface with mean orientation corresponding to some low index plane the monoatomic step array may be written as

$$\rho(x) = \rho_0 \cos(2\pi x/\lambda) \tag{16}$$

where $\rho(x)$ is the density of steps (number per unit length of surface) at x, ρ_0 its value at the points of inflection, with negative values of ρ taken to indicate steps of opposite sign. The problem of predicting the evolution of the surface shape then transforms to the problem of predicting the time dependence of $\rho(x)$. The sine wave amplitude decays by the annihilation of unlike steps at the maxima and minima. Two basic ingredients are required in order to proceed with the analysis of this problem. We need to know (i) the effective force (if any) on a step in terms of the local step distribution and (ii) the dominant mechanism by which atoms are transported between adjacent steps.

The step-step interaction potential is not yet well established for any surface. The interaction may be of long range elastic origin /23/ or could be effectively of long range due to short-range interaction at near contact points of neighboring steps that 'wander' close to one another. By rather indirect reasoning, BLAKELY and SCHWOEBEL /24/ have deduced that the average interaction potential is of the form $1/L^2$, where L is the spacing of the pair of steps. This conclusion was based on comparing the expressions for the deviation in defect density from that of a flat surface at a point where there is curvature K in terms of the Gibbs-Thomson relation and in terms of the net force on a step. A more fundamental derivation of the interaction would clearly be more desirable.

The principal mechanisms of atom transport between steps and of step annihilation during mass transport are really not well established, although very plausible descriptions can be given. Due to the net forces on steps the local equilibrium concentration of terrace defects (adatoms or vacancies) will differ from one point to another on the surface so that diffusive fluxes of those defects will be set up between steps. By acting as sources and sinks the steps release or absorb the defects and hence experience a net motion. The step motion could be limited either by diffusive transport across the terraces or by emission and incorporation of atoms at the steps; as indicated by several experiments /5/ the rate constants for emission and incorporation may depend on the direction from which an adatom approaches a step. It would be extremely helpful to have computer simulations of the step evolution processes. The parameters required for such a simulation would include step-step interaction potentials, kink formation energies, excess energy per unit length of step, and free energies of formation and motion of adatoms. Comparison of the computed surface changes with those observed after annealing should help to limit the range of the parameters involved and indicate which processes are of greatest importance. It should also be possible to simulate step clustering processes recently observed on several crystal surfaces /25,26,27,28/. The possibility of directly observing step motions on macroscopic crystals (and perhaps even adatom or vacancy diffusion) now exists through the development of the scanning tunneling microscope (STM) /29,30,31,32/, as well as through other techniques /33/. STM experiments on surfaces with sinusoidal step density modulations are now underway in the authors' laboratory. Related observations on surface mass transfer are also given in the interesting paper by BEHM in this volume.

References

1. R.W. Balluffi and J.M. Blakely: in Low Temperature Diffusion and Applications to Thin Films, ed. by A. Gangulee, P.S. Ho and K.N. Tu (Elsevier, Lausanne, 1975)

2. W.K. Burton, N. Cabrera, and F.C. Frank: Phil. Trans. Roy. Soc. London 243A, 299 (1951)

3. J.D. Weeks: in Ordering in Strongly Fluctuating Condensed Matter Systems, ed. by T. Riste (Plenum, New York, 1980)

4. G. Ehrlich and K. Stolt: Ann. Rev. Phys. Chem. 31, 603 (1980)

5. H.W. Fink and G. Ehrlich: J. Chem. Phys. 81, 4657 (1984)

6. E.H. Conrad, R.M. Aten, D.S. Kaufman, L.R. Allen, T. Engel, M. den Nijs and E.K. Reidel: J. Chem. Phys 84, 1015 (1986); J. Chem. Phys. 85, 4756 (1986)

7. J.W.M. Frenken and J.F. van der Veen: Phys. Rev. Lett. 54, 134 (1985)

8. J. Frenkel: in Kinetic Theory of Liquids (Oxford Univ. Press, Oxford, 1946)

9. K. Lehovec: J. Chem. Phys. 21, 1123 (1953)

10. J.M. Blakely and S. Danyluk: Surf. Sci. 40, 37 (1973)

11. C. Herring: in Physics of Powder Metallurgy, ed. by W.F. Kingston (McGraw-Hill, New York, 1951)

12. W.W. Mullins: in Metal Surfaces (Amer. Soc. for Metals, Metals Park, Ohio, 1963)

13. J.M. Blakely and H. Mykura: Acta Met. 10, 565 (1962)

14. P.S. Maiya and J.M. Blakely: Appl. Phys Letters 7, 60 (1965)

15. H.P. Bonzel: in Surface Mobilities on Solid Materials, ed. by V.T. Binh, NATO ASI series, Vol. 86, (Plenum, New York, 1983) p. 195

16. D. Makowiecki et. al.: (to be published)

17. F.H. Huang, R.A. Henrichsen and C.Y. Li: Materials Sci. Res. 10, 173 (1975)

18. C. Monty and J. LeDuigan: High Temperature-High Pressure 14, 709 (1982)

19. S. Mukhopadhyay et. al.: (to be published); see also A.P. Jardine et. al.: Mat. Res. Soc. Symp. Proc., Vol. 74, 365 (1987)

20. H.P. Bonzel and N.A. Gjostein: Appl. Phys. Letters 10, 258 (1967)

21. J.M. Blakely and D.L. Olson: J. Appl.Phys. 39, 3476 (1968)

22. M. Keefe: (to be published)

23. N. Cabrera: Surf. Sci. 2, 320 (1964)

24. J.M. Blakely and R.L. Schwoebel: Surf.Sci. 26, 321 (1971)

25. H.V. Thapliyal and J.M. Blakely: J.Vac. Sci. Technol. 15, 600 (1978)

26. B.Z. Olshanetsky and V.I. Mashanov: Surf.Sci. 111, 414 (1981)

27. F.A. List , P.S. Frankwicz and J.M. Blakely: Appl. Surf. Sci. 19, 161 (1984)

28. R.J. Phaneuf and E.D. Williams: Phys. Rev. Lett. 58, 2563 (1987)

29. G. Binnig and H. Rohrer: Physica B 127, 37 (1984)

30. R.S. Becker, J.A. Golovchenko, E.G. McRae and B.S. Swartzentruber: Phys. Rev. Lett. 55, 2028 (1985)

31. P.E. Wierenga, J.A. Kubby and J.E. Griffith: Phys. Rev. Lett. 59, 2169 (1987)

32. R.M. Feenstra and J.A. Stroscio: Phys. Rev. Lett. 59, 2173 (1987)

33. K. Yagi: in Proceedings of 2nd International Conference on the Structure of Surfaces (to be published)

The authors' work in this area is supported by the Materials Science Center at Cornell University, NSF grant #DMR-8403444, and by SRC.

Radiation Enhanced Diffusion (RED) in a Sputtered Ag/Ni Layered System

D. Marton, J. Fine, and G.P. Chambers*

Surface Science Divison, National Bureau of Standards, Gaithersburg, MD 20899, USA

1. INTRODUCTION

The layered Ag/Ni system has been shown to exhibit surface segregation of Ag when an Ag layer buried between two Ni layers is sputter profiled at room temperature [1]. While that study was directed to the analysis of radiation induced segregation (RIS), it showed also the role of diffusion in the transport of Ag atoms to the surface. The segregation rate was shown to obey first order kinetic equations, and segregation rates for 1 keV and 4 keV sputtering were determined.

Previous experimental studies of RED [2-9] have usually consisted of 3 steps: (1) a sample is prepared containing a marker of well-known depth distribution; (2) the sample is then irradiated (ion bombarded) and (3) the depth profile of the marker is obtained (this last step often includes the observation of RIS). Studies of RED caused by ion sputtering are complicated by the erosion of the sample during irradiation [9]. The depth profiles obtained in step 3 are often also obtained by sputter profiling (often at an energy much lower than that used in step 2) which, too, can lead to additional complexity in evaluation.

The aim of the present study has been the analysis of RED occurring during ion bombardment (thus combining steps 2 and 3) using Auger depth profiling. Multilayered Ag/Ni samples were sputter-depth profiled and the shape of the depth profiles of the Ag layers obtained under steady-state conditions were described using s a simple model, which includes mainly roughening and RED, the latter being characterized with an effective diffusion rate. This approach has the advantage that the results obtained are directly applicable to Auger (and SIMS) depth profiling, i.e., the contributions to interface broadening obtained in this study will have direct bearing on profiles obtained in AES (and SIMS) experiments. The results of this analysis will allow us to distinguish the contribution from each of these effects. Evaluation of Ni and Ag Auger depth profiles which take into account those separate contributions to interface broadening is rather complex and will be described in detail elsewhere [10]. This separation can be confirmed by an independent measurement of the light scattering from the surface, since the light scattering is directly related to surface roughness. We have carried out in-situ surface roughness measurements on the same samples and on the same surfaces studied by AES depth profiling. The light scattering results strongly support those results described in the present paper; further details will be published elsewhere [11].

---

*On leave from the Technical University Budapest, Hungary

2. EXPERIMENTS AND RESULTS

2.1 AES Experiments

Both RED and surface roughening which result from ion bombardment of the Ag/Ni system were analyzed using multilayered Ag/Ni thin film structures. These structures consist of five thin (3-4 nm) Ag layers alternately deposited between six thick (50 nm) Ni layers. Such samples were produced by sputter deposition using polished Si substrates that had been coated, in situ, with amorphous Si prior to the deposition of the metal films and are similar to those previously described [1]. Stylus measurements indicated that the r.m.s. surface roughness of the samples prior to sputtering was less than 1 nm.

Sputter-depth profiling of these Ag/Ni samples was carried out using Ar^+ ions of both 1 keV and 4 keV energy. A differentially pumped ion gun produced a fine focussed beam scanned over about 2x2 mm^2 at total beam currents in the 50 to 500 nA range; Auger analysis was performed using a high resolution single-pass cylindrical mirror analyzer described earlier [12]. The base pressure in the vacuum system during the experiments described here was typically $3x10^{-9}$ Torr. Auger electron-spectra excited by a primary electron beam of 2500 eV energy (50-200 nA) were recorded in a EN(E) mode; digital background subtraction techniques were employed to obtain high precision depth profiles. The elements detected were Ni (LMM line, 844 eV), Ag (MNN line, 349 eV) and nitrogen (KLL line, 375 eV).

2.2 Results

A typical AES depth profile is shown in Fig. 1. The abscissa is the sputtering time, which can be readily transformed to a depth scale using known layer thicknesses. Since it is not straightforward to deduce interface width, surface roughness and diffusion rate data from Auger-depth profiles, we return to the discussion of such results following the description of the evaluation procedure.

3. EVALUATION OF AUGER DEPTH PROFILES

The evaluation procedure for each Ni and Ag depth profile, shown in Fig. 1, is based on a series of assumptions and simplifications.

Each depth profile of a given individual Ag layer is analyzed independently. Each such depth profile resembles a Gaussian in shape but is asymmetric, having a steeper slope at earlier sputtering times. We propose that each of these Ag profiles is the result of the superposition of a Gaussian (symmetric) and a RED (asymmetric) profile. The width of the Gaussian is determined mainly by surface roughness, although a small contribution from other factors is not excluded. One such contribution is the thickness of the g layer, another is the escape depth of Auger electrons, a third is interface broadening due to cascade mixing, a fourth is that due to thermal diffusion. Since these factors are small and since there are no grounds to suggest that the surface roughness distribution is anything other than normal, our assumption is that all these combined factors yield a Gaussian depth profile shape:

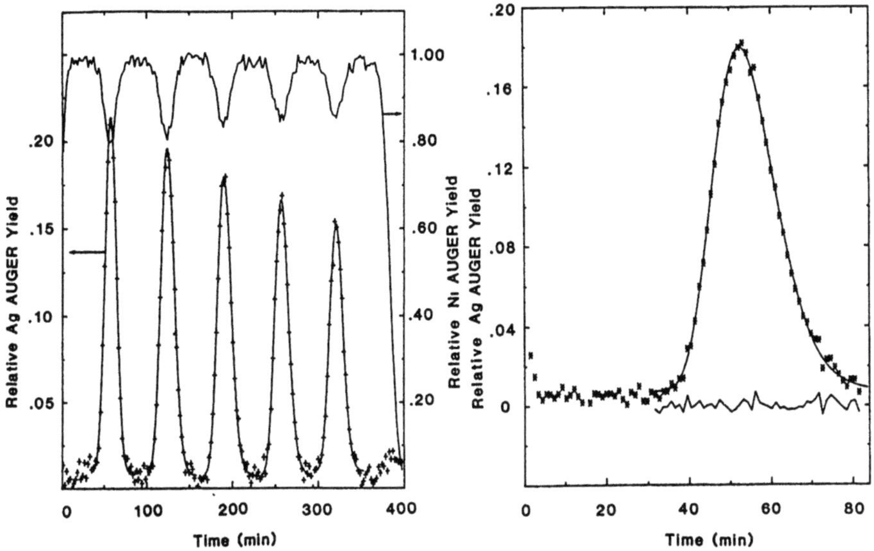

Fig. 1. Auger depth profile of a Ni/Ag multilayer (E_p = 1 keV). The left-side ordinate is the relative N(E) yield for Ag; the right-side ordinate is the same for Ni. The measured Ag data are shown as +, the Ni data are connected with straight lines. The full line through the Ag data represents a fit corresponding to eq. (5).

Fig. 2. Auger depth profile of an Ag layer (E_p = 4 keV). This, more detailed figure, is intended to show the quality of the fit. The measured points are *, the fitted curve is the full line, the difference between the measured and corresponding fitted points is shown as a jagged line (around the zero yield line).

$$c = (2^{1/2}c_o/\Delta z_R)\exp[-2(z-z_o)^2/\Delta z_R^2], \tag{1}$$

where c_o depends on the total number of the Ag atoms per surface area in a single silver layer, z is the depth coordinate, directed perpendicular to the surface, and z_o is the distance to the Ag layer, Δz_R^2, can be expressed as

$$\Delta z_R^2 = \Delta z_r^2 + d_{Ag}^2 + \ell_e^2 + \Delta z_c^2 + \Delta z_d^2, \tag{2}$$

in which Δz_r, Δz_c and Δz_d are the contributions of surface roughness, cascade mixing, and thermal diffusion respectively, and ℓ_e is the inelastic mean free path of Auger electrons. d_{Ag} stands for the thickness of the Ag layer.

Furthermore, we assume that RED starts when the sputtered surface is at a distance of z_o from the Ag layer, and that it continues at a constant rate after that. This will lead, according to standard diffusion theory for a planar source of negligible thickness at z_o, to a profile described by the equation [13]:

$$c = c_o(4Dt)^{-1/2}\exp[-(z-z_o)^2/4Dt], \tag{3}$$

where D is the (effective) diffusion coefficient and t is the diffusion time. Time t is related to depth z, and the sputtering rate S, according to the above definition for z_o by

$$z = St. \tag{4}$$

Note that time t for each Ag layer is measured separately.

Combining equations (1)-(4) we obtain an equation which describes the measured depth profiles:

$$c = c_o(4Dt+\Delta z_R^2/2)^{-1/2}\exp[-(St-z_o)^2(4Dt+\Delta z_R^2/2)^{-1}]. \tag{5}$$

Depth profiles, containing c vs t information, are used in a computer fitting program to calculate c_o, D and Δz_R, as well as a background for c. This equation can be used to obtain very precise fits to the asymmetric Ag-layer sputter depth profiles; an example of such a fit is shown in Fig. 2.

The model described allows us to calculate interface broadening values for each sputtered Ag layer. In this procedure we first calculate the maximal Ag concentration. Then the sputtering rate, S is evaluated, supposing that the sputtering yield of the Ag-Ni mix is the weighted average of the elemental sputtering yields of Ag and Ni. The variation of S within the Ag layers was neglected owing to the small variation of the Ag concentration in the Ag depth profiles. At this point we determine the thickness of the Ag layer, d_{Ag} and the distance between the depth immediately before the Ag Auger intensity (concentration) starts to increase and the unsputtered Ag layer itself. This distance is taken to be equal to z_o. This choice of z_o was justified by the results of the following experiments: when a depth profiling measurement is interrupted at some depth before the first Ag layer is detected, but the Ag Auger yield is recorded further, it is possible to observe two, dramatically different outcomes, i.e., Ag either segregates on the surface, or it does not segregate. In the first case the ion induced defects were able to reach the Ag layer, in the second case, however, the Ag layer is still further than the range of the defects. A series of such experiments is sufficient to determine the range of the defects with an accuracy of about 1 nm. In a forthcoming paper we will present detailed experimental results concerning z_o [10]. In the present study, we used values around 16 nm and 20 nm for z_o at 1 keV and 4 keV bombardment, respectively. It should be pointed out also that the diffusion rate data obtained due to the present treatment are quite insensitive to the value of z_o. Now, eq. (5) is used to fit the experimental depth profile and c_o, D and Δz_r is then obtained form eq. (5). Here ℓ_e = 0.8 nm (Ag at 375 eV [14] has been used for all measurements. Values for Δz_c for various sputtering ions and energies are available from calculations based on Monte-Carlo methods [15]. The values used here are 2 nm for E_p = 1 keV and 4 nm for E_p = 4 keV bombardment. The thermal interdiffusion in the Ag/Ni system is probably very small due to the low solubility of silver in nickel [16] The thermal diffusion contribution was also checked by carrying out some experiments at elevated temperatures. After a heat treatment of the sample in the vacuum at 330°C for 24h no change in the depth profiles could be detected. Thus Δz_d is neglected in the present calculations.

The full interface width obtained in our measurements consists of a part due to RED. This part can be characterized by the quantity

$$\Delta z_D = (8Dz_o/S)^{1/2}, \qquad (6)$$

the "diffusion interface width" which plays an analogous role to Δz_R in eq. (5) for $t=z_o/S$, i.e., for the time necessary to reach the Ag layer.

4. DISCUSSION

4.1 The Interface Broadening

Interface broadening $\Delta z=(\Delta z_R^2+\Delta z_D^2)^{1/2}$ and Δz_r both increase with the energy of the bombarding ions and with the sputtered depth. An example of results of measurements in which both AES depth profiling and light scattering was applied, is included in Table 1.

Table 1.

Energy [keV]	Layer/Depth	[nm]	Δz_r [nm] from AES	Δz_r [nm] from light scattering	Δz_D [nm]
1	1	50	7.4	4.9	6.0
	2	100	-	9.7	-
	3	150	9.3	11.8	7.3
	4	200	11.0	12.9	7.7
	5	250	13.6	13.8	7.3
4	1	50	11.9	10.1	14.8
	2	100	14.8	15.1	15.6
	3	150	18.0	17.4	13.2
	4	200	15.5	18.8	13.1
	5	250	19.8	19.4	9.8

The analysis of the data clearly shows that Δz_D is independent of the depth, and thus the depth dependence of the interface width is due to the roughening of the surface. This is as expected, since there is no reason, why the RED process should be affected by the depth, whereas the roughening of the surface is expected to increase as the square root of the depth [17].

4.2 The Diffusion Rates

Diffusion coefficients obtained from experimental depth profiles of Ag, using eq. (5) to fit the data, are plotted in Fig. 3 as a function of sputtering rates S_{Ni}. Results obtained with both 1 keV and 4 keV sputtering energies, E_p, are described by linear equations:

$$D = 1.22\times10^{-17} + 1.75\times10^{-15}S_{Ni} \quad (r=0.9710,\ n=43,\ E_p=1\ \text{keV}) \qquad (7a)$$

and

$$D = 1.35\times10^{-17} + 9.05\times10^{-15}S_{Ni} \quad (r=0.9764,\ n=29,\ E_p=4\ \text{keV}) \qquad (7b)$$

where D is in cm^2/s and S_{Ni} is in nm/s units; r is the regression coefficient and n is the number of data points taken into consideration.

In Fig. 3, data obtained in the same sputtering profile (constant sputtering rate) from a number of Ag layers at different depths are

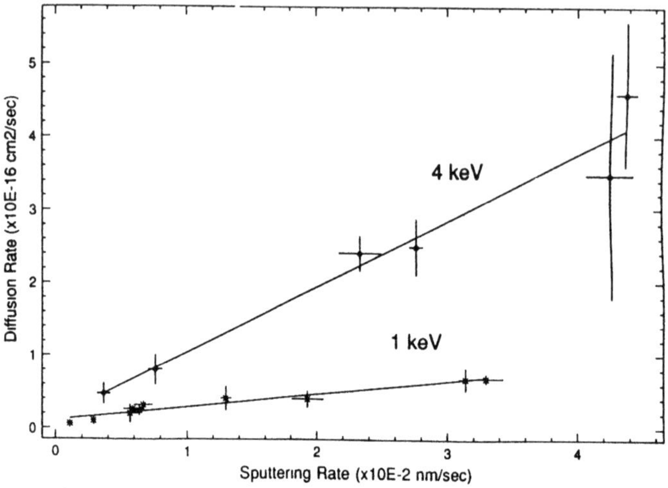

Fig. 3. RED rates obtained from AES depth profiling experiments vs. sputtering rate. Each point corresponds to one run, i.e., several (in most cases 5) data points. The error bars correspond to the standard deviation of these points. o - 1 keV, x - 4 keV energy of incident ions.

shown as a single point, with an error bar corresponding to the scatter of the sputtering rate and diffusion rate of the various layers. The diffusion rates are plotted against the sputtering rates rather than against current densities, as the latter are not easily measurable. Since sputtering rates are proportional to current densities for any given energy of bombarding ions, equations (7a,b) describe linear dependences of the measured diffusion rates on the ion beam current density. The non-zero D(0) values (which correspond to a linear extrapolation of S_{Ni} to zero, about $1.3x10^{-17}$ cm^2/s) are probably due to measurement errors, or due to asymmetry caused by factors other than RED. Nevertheless, we can state that the RED rates are essentially proportional to the sputtering rates. This is a very important result, because this is just what would be predicted from the theory developed by Dienes and Damask [18] for RED. Thus the linear relationship of D vs. S_{Ni} shown in Fig. 3 would suggest that the evaluation procedure followed in the present paper is consistent with the above theory.

The diffusion rate of the grain boundary diffusion of Ag in Ni, extrapolated from high temperature data [19] is about $D_b=3.3x10^{-19}cm^2/s$ at 300K, a value considerably smaller than what is obtained here for the RED rate. The authors are not aware of RED rate measurements carried out on the Ag/Ni system, except the RIS measurement cited above [1]. The diffusion rate was not evaluated in that study, but can be estimated on the basis of the segregation rate constants, which were obtained. This estimate yields $D_1=4.7x10^{-17}cm^2/s$ and $D_4=1.9x10^{-16}cm^2/s$ for 1 keV and 4 keV experiments, respectively. (The sputtering rates were $S_1=7x10^{-3}$ nm/s and S_4=3.1x10-2nm/s.) These D values obtained from a completely different measurement are in good agreement with those reported here.

Swartzfager et al [4] observed D values of the order of magnitude $10^{-16}cm^2/s$ in Cu-Ni, $10^{-17}cm^2/s$ in Au-Pd and $10^{-17}cm^2/s$ in Ag-Au alloys, using 2 keV Ne^+ sputtering at room temperature. The sputtering rate was 5.2×10^{-3} nm/s, comparable with some of our data. We observed diffusion rates surprisingly close to these data (cf. Fig. 3). Li and Koshikawa [20] obtained a diffusion rate around 1.1×10^{-14} cm^2/s for the Cu-Au system (2 keV Ar^+ bombardment) at room temperature. This high diffusion rate can, in part, be explained by the fact that the authors used a high current density (0.22nm/s erosion rate), nearly 5 times the maximum rate used in the present study.

4.3 The Defects

The theory of radiation enhanced diffusion in solids has been described by Dienes and Damask [18]. According to this theory, for the case where the defect generation rate is constant and the annealing rate of defects is linear (i.e., defects anneal after random walks to sinks), the rate of the radiation enhanced diffusion is

$$D = \frac{K f_{Ag} \lambda_{Ag}^2}{R f_d \lambda_d^2}, \qquad (8)$$

where K is a constant rate of defect concentration production, and R is a geometrical constant that depends on the location and shape of the sinks where defects annihilate; f_{Ag} and f_d are the jump frequencies of Ag atoms and defects, respectively, and λ_{Ag} and λ_d are the jump distances. This expression for the diffusion rate does not include any contribution from thermal processes which, at room temperature, are known to be quite small.

The defect production rate for a given ion bombardment energy should be proportional to the bombarding ion current density, in other words, to the number of primary ions incident onto the sample surface per unit area and unit time. We denote this quantity by v. This relation will hold if the defects are produced by ion bombardment and each ion impact can be considered as an individual event. In the present experiments, sufficiently low current densities were used ($\ll 1 nA/cm^2$) so that this requirement is fulfilled and indeed a linear relationship does exist between the diffusion rate observed and the incident ion beam current density. It must be emphasized that the linear relationship between the current density and the diffusion rate has not been studied systematically by previous authors, although in [20] there are some results suggesting such a relationship.

In the following discussion an estimate of the number of the defects created by each bombarding particle, necessary to produce the observed diffusion rate, will be given.

Very little reliable data exists for the sputtering yield of Ni at 1 keV Ar^+ bombardment, and no data is available at 4 keV. Therefore, the following calculations are of limited accuracy but are, we believe, good estimates. Taking $Y_{Ni,1}$ = 2.16 atoms/ions [21], for 1 keV and assuming that $Y_{Ni,4} = 2Y_{Ni,1}$ for 4 keV, one can calculate the conversion factors between the sputtering rates and the current densities in eq. (7).

If Y_d is the number of defects produced per impinging ion, then

$$K = Y_d v V_{at}/z_o \qquad (9)$$

where V_{at} is the atomic volume of the target atom. Combining eqns. (8) and (9)

$$Y_d = \frac{Dz_o R f_d \lambda_d^2}{v V_{at} f_{Ag} \lambda_{Ag}^2}, \tag{10}$$

where D/v, z_o, and V_{at} are now known quantities and R, f and λ depend on the material, i.e., on defect structure and annihilation mechanism.

The defect annihilation mechanism is not expected to depend on the energy and/or density of the primary ions provided that it does not involve the recombination of defects with each other. The linear relationship observed between the sputtering rates and the RED rates allows us to exclude this recombination mechanism. The very fact that diffusion takes place on a time scale of minutes (both in experiments described here and in [1], allows us to conclude that defect complexes, rather than simple point defects, are responsible for the diffusion observed. Such defect complexes would probably not be easily annealed on account of their binding energy. It would appear that the segregation time constants are significantly shorter than the annihilation times and this would imply that $f_d < f_{Ag}$. In order to obtain a numerical assessment of the number of defects produced by an impinging ion (Y_d), we will assume that $f_{Ag} = 2f_d$, while the jump distances are equal. Using our experimental data, the following values for Y_d are obtained: Y_{1d}=1.9 defects/ion (1 keV) and Y_{4d}=20.4 defects/ion (4 keV). These estimated Y_d values are based on the assumption that the Ag diffusion is related to the production of complex defects whose annealing time constants are long compared to that of the point defects; consequently point defects do not contribute to the observed diffusion rates. Although no experimental values exist for Y_d in Ag/Ni, results obtained by computer simulation techniques do exist for solids bombarded by keV ions and indicate that our values of Y_d are somewhat lower, but close [22,23]. One should be aware however that the computer simulation calculations are concerned primarily with the generation of simple point defects.

Our experimental values for the range of radiation damage which plays a role in RED (z_o) are much greater than the mean values of the depth distribution of defects obtained in computer simulation calculations. These calculations show, however, the existence of a long exponential tail in the defect distribution [22,23]. Such defects, lying in the tail of the defect distribution have also been experimentally detected using positron annihilation techniques in aluminum bombarded by low energy Ar^+ ions [23]. Another possible explanation of the large z_o values observed is the diffusion of some defects created in the collision cascade beyond this cascade, and so extend the range of radiation influence. This possibility was pointed out also by Ho [24].

5. SUMMARY

Radiation enhanced diffusion measurements have been carried out by Auger sputter depth profiling of Ni/Ag multilayered thin-film structures in order to determine the effect of room temperature RED on the interface broadening in sputter depth profile measurements. The interface broadening observed in this system is determined by a number of factors, the main two being roughening and RED. These two factors can be separated using the assumption that roughness leads to a symmetric profile shape, while and asymmetric shape is caused by RED. This concept is confirmed by interface broadening results which increase with depth (for subsequent Ag layers) for surface roughness, but remain independent of depth for diffusion.

Effective diffusion rates substantially higher than the grain boundary diffusivity of Ag in Ni at room temperature and proportional to the sputtering rates were determined for primary ion energies of 1 keV and 4 keV. The values obtained are in good agreement with RIS rates determined in previous experiments. Defect complexes, rather than point defects, are likely to play an important role in RED. It is important to note that the range of the RED process (i.e., the range of the defects responsible for it) is considerably greater than the range of ion mixing.

Beyond the particular findings concerning the Ni/Ag system, our study suggests that RED contribute substantially to interface broadening in some multicomponent systems which have relatively low activation energies of diffusion.

6. ACKNOWLEDGMENTS

Authors would like to thank B. Navinsek (J. Stefan Inst., Yugoslavia) who prepared the samples, C. J. Powell, C. Handwerker, T. Vorburger, (NBS), P. H. Holloway (Florida State Univ.) and T. D. Andreadis (NRL) for helpful discussions. One of us (D.M.) would like to express his gratitude to the Hungarian and U.S. Governments for their support and Fulbright stipend.

7. REFERENCES

1. J. Fine, T.D. Andreadis, F. Davarya: Nucl. Instr. & Methods, 209, 521 (1983)
2. R.R. Hart, H.L. Dunlap, O.J. Marsh: J. Appl. Phys., 46, 1947 (1975).
3. J.E. Hobbs, A.D. Marwick: Radiation Eff. Lett., 58, 83 (1981).
4. D.G. Swartzfager, S.B. Ziemecki, M.J. Kelley: J. Vac. Sci. Technol. 19, 185 (1981)
5. J.E. Hobbs, A.D. Marwick: Nucl. Instr. & Meth. in Phys. Res., B9 169 (1985)
6. L. Kornbilt, A.R. Zomorrodian, S. Tougaard, A. Ignatiev: Rad. Effects, 91, 97 (1985)
7. W. Vandervorst, F.R. Shepherd, M.L. Swanson, H.H. Plattner, O.M. Westcott, I.V. Mitchell: Nucl. Instr. & Meth. in Phys. Res., B15, 201 (1986)
8. D. Farkas, R. Pasianot, M. Rangaswamy, E.J. Savino: Nucl. Instr. & Meth. in Phys. Res., B16, 183 (1986)
9. R. Collins: In 3rd Conf. on Radiation Effects in Insulators, Guilford, Surrey, England, July, 1985
10. D. Marton, J. Fine, G.P. Chambers: in preparation
11. D. Marton and J. Fine: in preparation
12. J. Fine, B. Navinsek: J. Vac. Sci. & Technol., A3, 1408 (1985)
13. P.G. Shewmon: Diffusion in Solids, McGraw-Hill, New York, 1963
14. J.C. Ashley, C.J. Tung: Surface & Interf. Anal., 4, 52 (1982)
15. F. Davarya, M.L. Rousch, J. Fine, T.D. Andreadis, and O.F. Goktepe: J. Vac. Sci. Technol., A1, 467 (1983)

16. Constitution of Binary Alloys, ed. M. Hansen, 2nd ed. K. Anderko, McGraw-Hill, New York, 1958
17. D. Marton and J. Fine: Thin Solid Films, 151, 433 (1987)
18. G.J. Dienes and A.C. Damask: J. Appl. Phys., 29, 1713 (1958)
19. A.B. Vladimirov, V.N. Kaygorodov, S.M. Klotsman, I. Sh. Trakhtenberg: Fiz. Metal. Metalloved., 45, 1015 (1978); (English transl.: Phys. Met. Metall., 45, 100 (1979)
20. R.S. Li and T. Koshikawa: Surface Sci., 151, 459 (1985)
21. H. Oechsner: Appl. Phys., 8, 185 (1975)
22. D.S. Karpuzov and D.G. Armour: J. Phys., D17, 853 (1984)
23. A. Vehanen, J. Makinen, P. Hautojarvi, H. Huomo, J. Lahtinen, R.M. Niemine, and S. Valkealahti: Phys. Rev., B32, 7561 (1985)
24. P.S. Ho: Surface Sci., 72, 253 (1978)

The Influence of Segregation on the Formation of Cu/Ni Alloy Films by Means of Evaporation

B. Hofmann, D. Borgmann, and G. Wedler

Institut für Physikalische und Theoretisch Chemie
der Universität Erlangen-Nürnberg, Egerlandstr. 3,
D-8520 Erlangen, Fed. Rep. of Germany

1. Introduction

During the last twenty years lots of papers have been published dealing with the surface segregation of Cu/Ni alloys /1-77/. The great importance of bimetallic catalysts as Cu/Ni alloys justifies this huge number of experimental /1-49/ and theoretical /50-77/ investigations. Reactivity and selectivity of alloy catalysts in heterogeneously catalyzed reactions are much higher than those of the pure metals. The composition of the surface is the decisive entity since it is responsible for the chemical and catalytic properties of the Cu/Ni alloys.

In the past we investigated adsorption effects and catalytic reactions of small molecules (e.g. H_2,O_2,N_2,CO,CO_2,C_2H_2) on evaporated films of pure metals (e.g. Fe,Ni,Cu,Pd) /e.g. 78-88/. In contrast to single crystals there is the possibility to change the structure and thereby the catalytic activity of the films over a wide range by changing the annealing conditions /e.g. 78,84,86,87/. Up to now a lot of information concerning the properties of pure metal films as adsorbents and catalysts is available. Additionally it would be interesting to measure the influence of alloying on the structural and electronic properties.

It is the aim of this investigation to study whether it is possible to produce Cu/Ni alloy films of a predetermined surface composition by means of successive evaporation of Cu and Ni films (or vice versa) onto a glass substrate under ultrahigh vacuum conditions and by annealing at temperatures not exceeding 600 K.

2. Experimental

The experiments were carried out in an all-metal uhv system pumped by a combination of an oil-diffusion and a Ti sublimation pump. After baking out the residual gas pressure was smaller than 1.3×10^{-8} Pa.

The recipient was equipped with
a manipulator carrying an insulated, cubic sample holder which could be heated by means of a well shielded bifilar tungsten coil allowing to keep the temperature constant at ±3 K (Fig. 1),
an electron gun for grazing incidence (75° against the normal) of the electron beam onto the sample

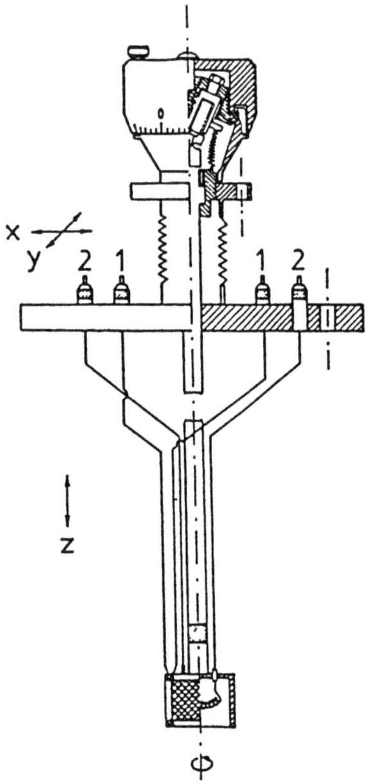

Fig. 1: Manipulator with cubic sample holder: 1 Pt/Pt-Rh (10%) thermocouple; 2 substrate heating with bifilar tungsten coils.

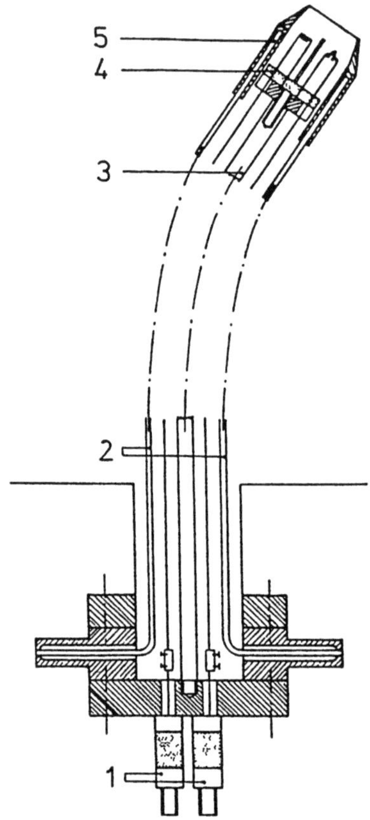

Fig. 2: Evaporation source with liquid nitrogen cooling shield: 1 current feedthroughs; 2 liquid nitrogen tubes; 3 source holder (grounded); 4 ceramic plate; 5 cooling shield.

a cylindrical mirror analyzer for AES,
an Ar^+-ion gun for sputtering,
an evaporation source (Fig. 2), and
a water-cooled quartz crystal thickness controller.

Cover-glasses with the dimensions 12 mm x 12 mm x 0.15 mm were used as supports for the films. They were pressed onto the sample holder by means of plate-springs. The temperature was measured by means of a Pt-Pt/10 % Rh thermocouple attached to the glass substrate.

Figure 2 shows the evaporation source. It is fitted to a CF 35 flange and consists of the holder for the evaporation coils (Ni 4.5 (MRC) and W bearing a drop of Cu 5.0 (MRC)) and a liquid nitrogen cooled shield which prevents a marked increase in pressure during evaporation.

The thickness of the metal films could continuously be measured during evaporation by means of the quartz crystal thickness controller. The exact determination of the thickness was performed after the experiment. The alloy films were solved in nitric acid (4.4 %) and then analyzed in an atom absorption spectrometer.

Cu films showed traces of contamination by sulfur, Ni films were sometimes contamined by carbon stemming from carbon monoxide. Such Ni films could be cleaned by sputtering when the Ni film was deposited before Cu was evaporated.

In order to form a Cu/Ni alloy the films were annealed at higher temperatures which, however did not exceed 583 K. During annealing Auger electron spectra were recorded. The annealing process was stopped when the Auger electron spectrum did not indicate any further changes with time. Figure 3 is the Auger electron spectrum of an annealed alloy film. The annealing temperature was 563 K, the annealing time 223 min. There is only a very small contamination by chlorine and carbon.

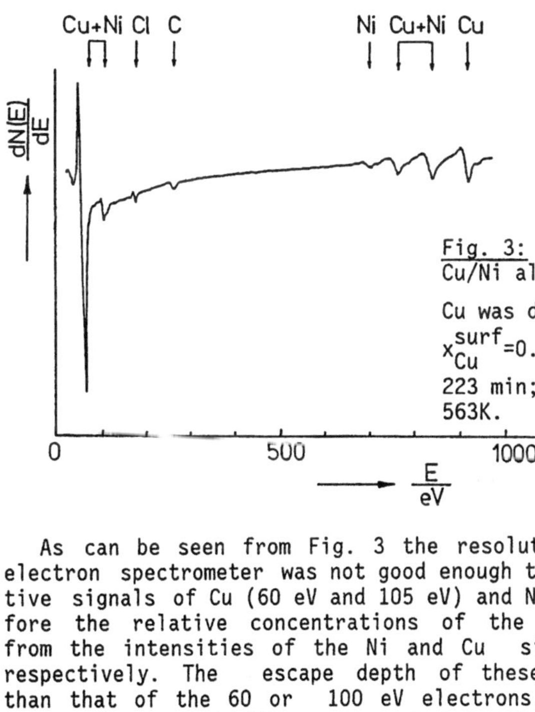

Fig. 3: Auger electron spectrum of a Cu/Ni alloy surface after annealing. Cu was deposited on Ni. x_{Cu}^{vol}=0.56, x_{Cu}^{surf}=0.68, annealing time t_a= 223 min; annealing temperature T_a= 563K.

As can be seen from Fig. 3 the resolution of the available Auger electron spectrometer was not good enough to resolve the surface sensitive signals of Cu (60 eV and 105 eV) and Ni (61 eV and 109 eV). Therefore the relative concentrations of the two metals were calculated from the intensities of the Ni and Cu signals at 716 eV and 920 eV, respectively. The escape depth of these Auger electrons is larger than that of the 60 or 100 eV electrons. However, this disadvantage is, at least partially, compensated by the application of the grazing incidence of the primary electrons.

The relative amounts of Cu and Ni were determined by means of the formula

$$x_A = \left[1 + \frac{I_B}{I_A} \cdot P_{rel}\right]^{-1} \qquad (1)$$

proposed by HALL et al. /89/. x_i is the mole fraction of component i, I_i the peak-to-peak height of the signal of element i, and P_{rel} the relative sensitivity factor defined by

$$P_{rel}(A,B) = \frac{I_A}{x_A} : \frac{I_B}{x_B} \qquad (2)$$

In spite of the fact that these sensitivity factors are compiled in literature /90/ they were derived in each experiment from the spectra of the pure Cu and Ni films.

Figure 4 shows how the mole fraction of Cu decreases with time when the interdiffusion is started by raising the temperature to 563 K. In case a) a Cu film of 14 nm thickness covered a Ni film of 7 nm thickness; in case b) the corresponding values were 5 nm and 10 nm.

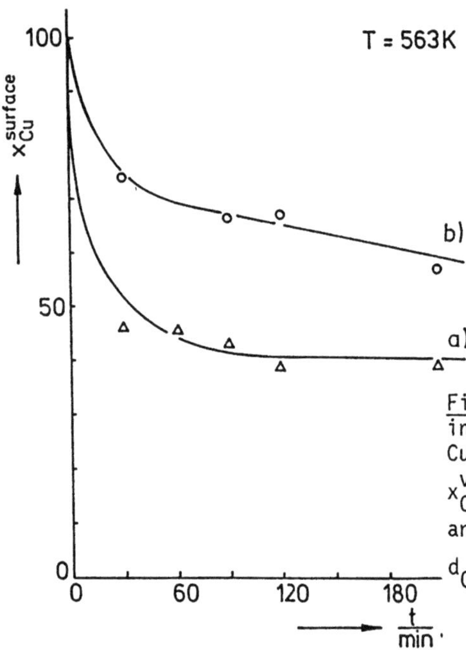

Fig. 4: Change in Cu concentration in dependence on annealing time. Cu was deposited on Ni. Curve a) x_{Cu}^{vol}=0.69, film thickness d_{Cu}=14 nm and d_{Ni}=7 nm; curve b) x_{Cu}^{vol}=0.33, d_{Cu}=5 nm, d_{Ni}=10 nm.

3. Results and Discussion

A large number of experiments was carried out in which the sequence of the deposition of the two metals, the relative amounts of Cu and Ni, the thickness of the outer layer, and the annealing temperature were systematically varied.

Both the experimental conditions and the results are compiled in Table 1. Figure 5 gives a survey of some of these data, which seem to be of particular interest: The upper lines demonstrate the

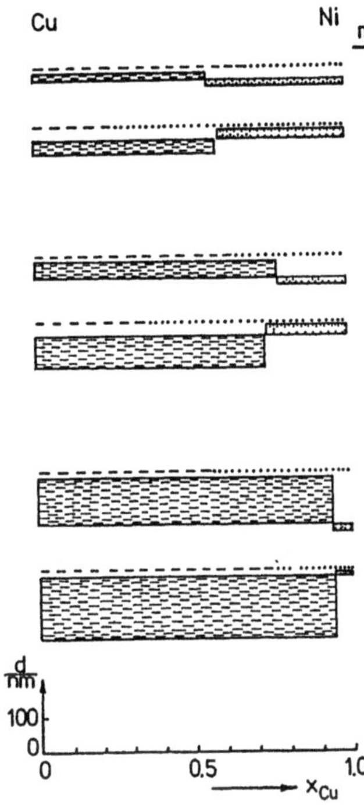

met.	$x^{vol.}$	$x^{surf.}$	d/nm
Cu	0.56	0.68	14
Ni	0.44	0.32	11
Ni	0.42	0.74	26
Cu	0.58	0.26	35
Cu	0.78	0.64	55
Ni	0.22	0.36	14
Ni	0.26	0.64	34
Cu	0.74	0.36	95
Cu	0.94	0.55	181
Ni	0.06	0.45	12
Ni	0.05	0.25	11
Cu	0.95	0.75	190

Fig. 5: Some selected experimental results. The plots show the bulk concentrations, film thickness and sequence of deposition (large rectangles) and the surface concentrations after annealing (thin lines).

Table 1: Experimental conditions and results. Annealing time t_a, annealing temperature T_a, film thickness d, mole fraction x, symbol * marks the top layer.

t_a/min	T_a/K	d_{Cu}/nm	d_{Ni}/nm	x_{Cu}^{vol}	x_{Ni}^{vol}	x_{Cu}^{surf}	x_{Ni}^{surf}
223	563	14*	11	0.56	0.44	0.68	0.32
210	563	10*	9	0.52	0.48	0.57	0.43
308	573	6*	11	0.36	0.64	0.65	0.34
87	583	55*	14	0.78	0.22	0.64	0.36
398	523	45*	21	0.69	0.31	0.65	0.35
260	553	181*	12	0.94	0.06	0.55	0.45
210	563	5*	10	0.33	0.67	0.40	0.60
210	563	14*	7	0.69	0.31	0.62	0.38
308	573	7*	13	0.37	0.63	0.51	0.49
308	573	7*	4	0.63	0.37	0.68	0.32
292	557	35	26*	0.58	0.42	0.26	0.74
318	527	26	10*	0.73	0.27	0.18	0.82
63	560	95	34*	0.74	0.26	0.36	0.64
191	553	190	11*	0.95	0.05	0.75	0.25
204	573	119	4*	0.97	0.03	0.57	0.43
216	516	321	6*	0.98	0.02	0.58	0.42
200	573	491	70*	0.88	0.12	0	1.0

composition of the surface after annealing (dashed line: Cu, dotted line: Ni). The thickness of the beams representing the upper and lower layer is proportional to the thickness of the layers, their lengths indicate the bulk composition. This composition is expressed as mole fraction

$$x_i = \frac{N_i}{N_i + N_j} \qquad (3)$$

where N_i and N_j are the numbers of atoms i and j, respectively, which are contained in the two layers i and j.

It is evident that under the given conditions

i) the metal that initially formed the outer layer remains dominant in the surface ($x > 0.5$)

ii) in many cases the Ni content in the surface surmounts that in the bulk. This is independent on whether Ni or Cu initially formed the upper layer.

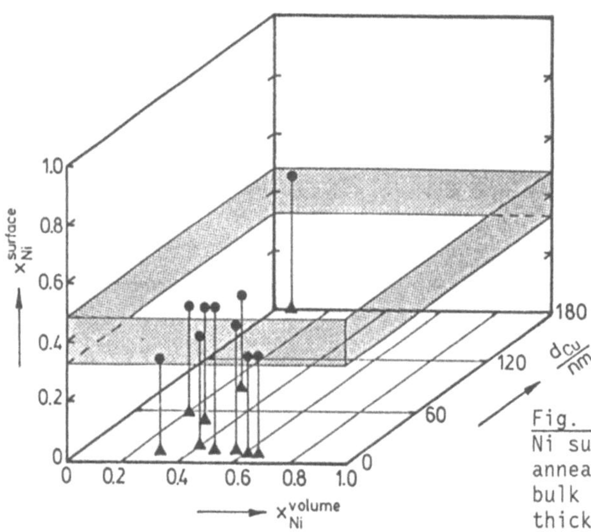

Fig. 6: 3-dimensional plot of Ni surface concentration after annealing in dependence on Ni bulk concentration and the thickness of the Cu top layer.

Figure 6 relates only to the experiments in which at first a Ni layer was deposited. Then these layers were covered with Cu layers of varying thicknesses. The triangles indicate the mole fraction of Ni in the bulk and the thickness of the covering Cu film. The full circles show the mole fraction of Ni in the surface. The surface composition is not markedly influenced by x_{Ni}^{volume} or d_{Cu}. $x_{Ni}^{surface}$ varies between 0.3 and 0.5 as it is indicated by the shadowed region.

Quite another observation is made when Cu films are covered by Ni layers (Fig. 7). In this case the composition of the surface strongly

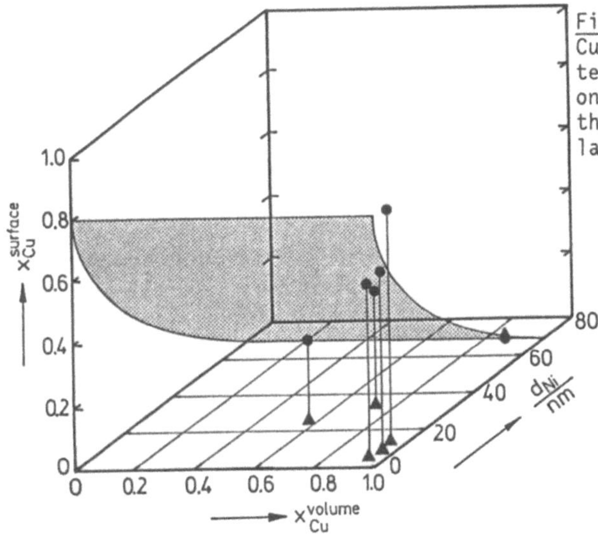

Fig. 7: 3-dimensional plot of Cu surface concentration after annealing in dependence on Cu bulk concentration and the thickness of the Ni top layer.

depends on the thickness of the coating Ni layer. Under the given experimental conditions a Ni film of 70 nm thickness seems to form a complete diffusion barrier for Cu.

The experimental procedure used in this work considerably differs from that applied by other authors who are particularly interested in elucidating the segregation of Cu or Ni starting with Cu/Ni alloys of various compositions. We started with a system consisting of a pure Ni (Cu) layer coated with a pure Cu (Ni) film.

In order to be able to compare our results with literature data the results given in Figs. 6 and 7 are plotted in a diagram mole fraction in the surface against mole fraction in the bulk. When doing this (Fig. 8) the important information regarding the thickness of the top layer is lost. The full circles refer to the experiments in which a Ni layer was coated by a Cu layer, the crosses refer to Cu layers coated by Ni.

Figure 9 shows corresponding data obtained by other authors using low energy ion scattering /25/, AES /19/ and atom probe technique /28/. They all find Cu segregation which is in agreement with theoretical predictions /51,52,57,60,62,63,67,68,74,75/.

On the other hand a theoretical work of KERKER et al. /53/ exists predicting a large Ni segregation in Cu rich alloys. Just recently SAKURAI et al./40,41,43,45,46,48/ reinvestigated the segregation in the Cu/Ni alloy system by means of the atom probe method. They also included Cu rich alloys with a bulk mole fraction exceeding 0.84 (Fig. 10). Their results confirm the Cu segregation in the range $0 \leq x_{Cu}^{volume} \leq 0.84$. For higher mole fractions, however, they observe a segregation of Ni. These results are supported by measurements of the catalytic activity of Cu/Ni alloys reported by SINFELT et al. /12/

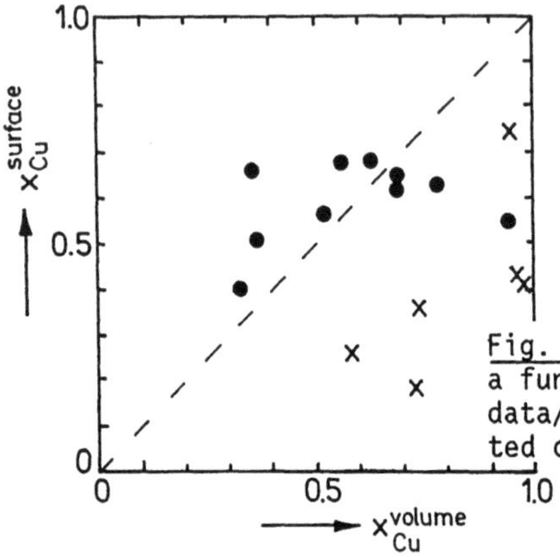

Fig. 8: Cu concentration at the surface as a function of Cu bulk concentration /own data/. ● Cu deposited on Ni, x Ni deposited on Cu.

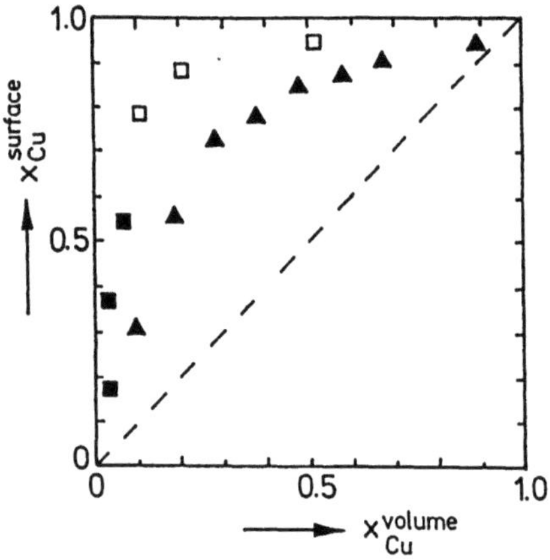

Fig. 9: Cu surface concentration vs. Cu bulk concentration. □ LEIS data /25/, ▲ AES data /19/ and ■ atom probe data /28/.

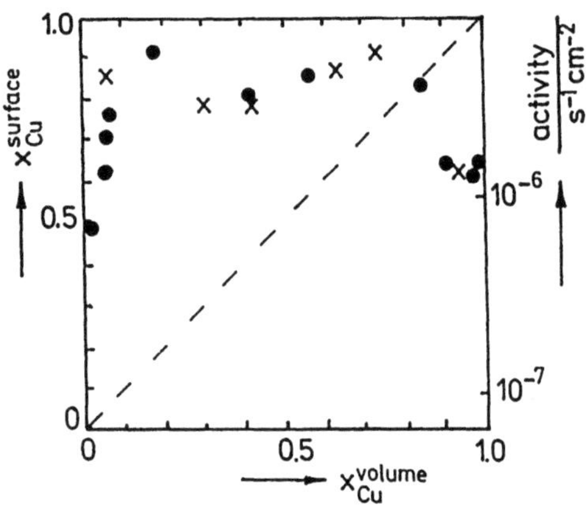

Fig.10: Cu surface concentration vs. Cu bulk concentration. ● atom probe data /48/, x catalytic activity data /12/.

who studied the dehydrogenation of cyclohexane under formation of benzene. KELLER /47/ performed SIMS measurements at polycrystalline Cu samples which indicate a segregation of Ni on the surface after annealing at 500 K for 20 hours.

Comparison of Figs. 6 to 8 with Figs. 9 and 10 clearly shows that in our case there is no marked Cu segregation. Ni is able to diffuse through a Cu coating to the surface (Figs. 6 and 8 (full circles)) until the mole fraction of Cu is lowered to $0.5 \leq x_{Cu}^{surface} \leq 0.7$.
This finally leads to a Ni segregation for values of $x_{Cu}^{volume} > 0.6$.
On the other hand diffusion of Cu through a Ni coating is strongly hindered (Fig. 7 and 8 (crosses)). Thicker Ni layers act as diffusion barriers for Cu which is already known from literature /91/.

This is true in spite of the fact that the activation energy of Cu in Ni (E_{act} = 126.4 kJ mol^{-1}) is smaller than that of Ni in Cu (E_{act} = 141 kJ mol^{-1}) /91/.

These observations prove that formation of an alloy took place in any cases where Ni was coated with Cu. The differences to literature data must be traced back

i) to the fact that the original material did not consist of an alloy but of two separate metal films

ii) to the application of much smaller annealing temperatures.

Up to now it is not yet possible to decide whether there is also an influence of the glass substrate and of diffusion along grain boundaries.

4. Conclusions

The experiments have shown that it is possible:

i) to produce Cu/Ni alloy films with surface compositions spanning a wide range, with rather high reproducibility for $0.5 \leq x_{Cu}^{surface} \leq 0.7$

ii) to avoid the strong segregation of Cu reported in the literature for higher annealing temperatures

iii) to observe a segregation of Ni

iv) to use a Ni layer as a diffusion barrier for Cu.

5. Acknowledgement

The authors thank the Deutsche Forschungsgemeinschaft and the Fonds der Chemischen Industrie for financial support.

1. W.M.H. Sachtler and G.J.H. Dorgelo: J. Catal. 4, 654 (1965)
2. W.M.H. Sachtler and R. Jongepier: J. Catal. 4, 665 (1965)
3. W.M.H. Sachtler, G.J.H. Dorgelo and R. Jongepier: J. Catal. 4, L 100 (1965)
4. P. van der Plank and W.M.H. Sachtler: J. Catal. 7, 300 (1967)
5. P. van der Plank and W.M.H. Sachtler: J. Catal. 12, 35 (1968)
6. L.A. Haris: J. Appl. Phys. 39, 1419 (1968)
7. W.M.H. Sachtler and P. van der Plank: Surf. Sci. 18, 62 (1969)
8. G. Ertl and J. Küppers: Surf. Sci. 24, 104 (1971)
9. D.T. Quinto, V.S. Surdaram and W.P. Robertson: Surf. Sci. 28, 504 (1971)
10. K. Nakayama, M. Ono and H. Shimizu: J. Vac. Sci. Technol. 9, 749 (1971)
11. G. Ertl and J. Küppers: J. Vac. Sci. Technol. 9, 829 (1972)
12. J.H. Sinfelt, J.L. Carter and D.J.C. Yates: J. Catal. 24, 280 (1972)

13. Y. Takasu and H. Shimizu: J. Catal. 29, 479 (1973)
14. C.R. Helms: J. Catal. 36, 114 (1975)
15. C.R. Helms and K.Y. Yu: J. Vac. Sci. Technol. 12, 276 (1975)
16. C.R. Helms, K.Y. Yu and W.E. Spicer: Surf. Sci. 52, 217 (1975)
17. H.H. Brongersma and T.M. Buck: Surf. Sci. 53, 649 (1975)
18. N.J. Shevchik and C.M. Penchina: Phys. Stat. Sol. B70, 619 (1975)
19. K. Watanabe, H. Hashiba and T. Yamashina: Surf. Sci. 61, 483 (1976)
20. Y. Takasu, H. Shimizu, S. Maru and Y. Matsuda: Surf. Sci. 61, 279 (1976)
21. K.Y. Yu, C.R. Helms and W.E. Spicer: Sol. State Comm. 18, 1365 (1976)
22. K.Y. Yu, C.R. Helms, W.E. Spicer and P.W. Chye: Phys. Rev. B15, 1629 (1977)
23. K. Watanabe, M. Hashiba, Y. Fukuda and T. Yamashina: Bull. Fac. Eng., Hokkaido University 83, 139 (1977)
24. J. Vrijen and S. Radelaar: Phys. Rev. B17, 409 (1978)
25. H.H. Brongersma, M.J. Sparnaay and T.M. Buck: Surf. Sci. 71, 675 (1978)
26. D.T. Ling, J.N. Miller, I. Lindau, W.E. Spicer and P.M. Stefan: Surf. Sci. 74, 612 (1978)
27. Y.S. Ng, T.T. Tsong and S.B. McLane: Phys. Rev. Lett. 42, 588 (1979)
29. T. Koshikawa, T. Goto, N. Saeki, R. Shimizu and E. Sugata: Surf. Sci. 79, 461 (1979)
30. K. Goto, T. Koshikawa, K. Ishikawa and R. Shimizu: Surf. Sci. 79, 461 (1979)
31. P.R. Webber and D. Chadwick: Surf. Sci. 94, L151 (1980)
32. T.T. Tsong, Y.S. Ng and S.B. McLane: J. Appl. Phys. 51, 6189 (1980)
33. M.J. Kelley, and V. Ponec: Progr. Surf. Sci. 11, 139 (1981)
34. P.R. Webber, C.E. Rojas, P.J. Dobson and D. Chadwick: Surf. Sci. 105, 20 (1981)
35. F. Toyokawa, K. Furuya and T. Kikuchi: Surf. Sci. 110, 329 (1981)
36. K.C. Khulbe and R.S. Mann: Catal. Rev. Sci. Eng. 24, 311 (1982)
37. P.J. Durham, R.G.Jordan, G.S. Sohal and L.T. Wille: Phys. Rev. Lett. 53, 2038 (1984)
38. J.R. Chelikowsky: Surf. Sci. 139, L197 (1984)
39. C.M. Loxton and J.S.T. Tsong: Surf. Sci. 139, 453 (1984)
40. T. Sakurai, T. Hashizume and A. Jimbo: Appl. Phys. Lett. 44, 38 (1984)
41. T. Hashizume, A. Jimbo and T. Sakurai: J. Physique 45, 435 (1984)
42. M. Ahmad and T.T. Tsong: Appl. Phys. Lett. 44, 40 (1984)
43. T. Sakurai, T. Hashizume, A. Jimbo, A. Sakai and S. Hyodo: Phys. Rev. Lett. 55, 514 (1985)
44. T. Ito, T. Shibanuma and T. Kadowaki: Jpn. J. Appl. Phys. 24, 144 (1985)
45. T. Sakurai, T. Hashizume, A. Jimbo, A. Sakai and S. Hyodo: Tech. Rep. of ISSP 1507A (University of Tokyo, Japan, 1985)
46. T. Hashizume, A. Jimbo and T. Sakurai: J. Vac. Sci. Technol. A3, 818 (1985)
47. G. Keller: Diploma work (Kaiserslautern, 1986)
48. T. Sakurai, T. Hashizume, A. Kobayashi, A. Sakai, S. Hyodo, Y. Kuk and H.W. Pickering: Phys. Rev. B34, 8379 (1986)
49. H.H. Brongersma, P.A.J. Ackermans and A.D. van Langeveld: Phys. Rev. B34, 5974 (1986)
50. G.M. Stocks, R.W. Williams and J.S. Faulkner: Phys. Rev. B4, 4390 (1971)
51. F.L. Williams and D. Nason: Surf. Sci. 45, 377 (1974)
52. J.J. Burton and E.S. Machlin: Phys. Rev. Lett. 37, 1433 (1976)

53. G. Kerker, J.L. Moran-Lopez and K.H. Bennemann: Phys. Rev. B15, 638 (1977)
54. C. Lea and M.P. Seah: Philos. Mag. 35, 213 (1977)
55. C.R. Helms: Surf. Sci. 69, 689 (1977)
56. J.L. Moran-Lopez and L.M. Falicov: Phys. Rev. B18, 2549 (1978)
57. A.R. Miedema: Z. Metallkunde 69, 455 (1978)
58. R.G. Donelly and T.S. King: Surf. Sci. 74, 89 (1978)
59. A. Crucq, L. Degols, G. Lienard and A. Frennet: Surf. Sci. 80, 78 (1979)
60. J.C. Hamilton: Phys. Rev. Lett. 42, 989 (1979)
61. F.F. Abraham, N.H. Tsai and G.M. Pound: Surf. Sci. 83, 406 (1979)
62. Ph. Lambin and J.P. Gaspard: J. Phys. F10, 2413 (1980)
63. F.F. Abraham: Phys. Rev. Lett. 46, 546 (1981)
64. V. Kumar: Phys. Rev. B23, 3756 (1981)
65. F.F. Abraham and C.R. Brundle: J. Vac. Sci. Technol. 18, 506 (1981)
66. K. Wandelt and C.R. Brundle: Phys. Rev. Lett. 46, 1529 (1981)
67. S. Mukherjee, J.L. Moran-Lopez, V. Kumar and K.H. Bennemann: Phys. Rev. B25, 730 (1982)
68. J.P. Muscat: J. Phys. C15, 867 (1982)
69. D. Tomanek, S. Mukherjee, V. Kumar and K.H. Bennemann: Surf. Sci. 114, 11 (1982)
70. R. Riedinger and H. Dreysse: Phys. Rev. B27, 2073 (1983)
71. W.F. Egelhoff, Jr.: Phys. Rev. Lett. 50, 587 (1983)
72. T.S. King and R.G. Donelly: Surf. Sci. 141, 417 (1984)
73. P.A. Dowben: Phys. Rev. B30, 7278 (1984)
74. O.L.J. Gijzeman: Surf. Sci. 150, 1 (1985)
75. H. Yamauchi: Phys. Rev. B31, 7688 (1985)
76. C.D. Stinesprung and W.F. Lawson: Surf. Sci. 150, 209 (1985)
77. U. Vahalia, P. Dowben and A. Miller: J. Vac. Sci. Technol. A4, 1675 (1986)
78. G. Wedler: Chemisorption (Butterworths, London, Boston 1976)
79. G. Wedler and J. Ertel: Ber. Bunsenges. Physik. Chem. 87, 469 (1983)
80. G. Wedler and M. Mengel: Surf. Sci. 131, L 423 (1983)
81. C. Pluntke, G. Wedler and G. Rau: Surf. Sci. 134, 145 (1983)
82. H. Körner, H. Landes, G. Wedler and H.J. Kreuzer: In Proc. 8th Intern. Congr. on Catalysis, Berlin 1984, Vol. IV, p. 145 (Verlag Chemie, Weinheim 1984)
83. H. Landes, A. Schmidt, G. Wedler and H.J. Kreuzer: Z. Physik. Chem. Neue Folge 144, 231 (1985)
84. G. Wedler, D. Borgmann, W. Alter and K. Witan: Ber. Bunsenges. Physik. Chem. 90, 235 (1986)
85. D. Borgmann, H. Behner, M. Pirner and G. Wedler: Langmuir 3, 261 (1986)
86. R. Bauer, H. Behner, D. Borgmann, M. Pirner, W. Spiess and G. Wedler: J. Vac. Sci. Technol. A5, 1110 (1987)
87. M. Pirner, R. Bauer, D. Borgmann and G. Wedler: Surf. Sci. 189/190, 147 (1987)
88. D. Borgmann, F. Kroninger, P. Steidl and G. Wedler: Surf. Sci. 189/190, 485 (1987)
89. P.M. Hall, J.M. Morabito and D.K. Conley: Surf. Sci. 62, 1 (1977)
90. P.W. Palmberg, G.E. Riach, R.G. Weber and N.C. Mac Donald: Handbook of Auger Electron Spectroscopy (Phys. Electr. Ind. Inc., Edina, Minn. 1976)
91. H. Lefakis, J.F. Cain and P.S. Ho: Thin Solid Films 101, 201 (1983)

Surface Properites of a Lead-Tin Alloy

W.N. Unertl, M. Grunze, and J.J. Weimer+

Laboratory for Surface Science and Technology,
University of Maine, Orono, ME 04469, USA

This report describes the surface properties of a lead-tin alloy with seventy weight percent lead. The equilibrium properties of clean surfaces are characterized for temperatures up to the bulk liquidus line. These results are discussed in the context of a theory of the variation of alloy surface composition during melting. We also studied the effects of ion bombardment and previous thermal history on the approach to equilibrium.

1. General Behavior During Binary Alloy Melting

The way in which the surface composition of a binary alloy changes during melting can be described most easily in terms of the bulk phase diagram. We couch this discussion in terms of the lead-tin phase diagram [1], Fig. 1, but the arguments can easily be extended to other alloy systems.

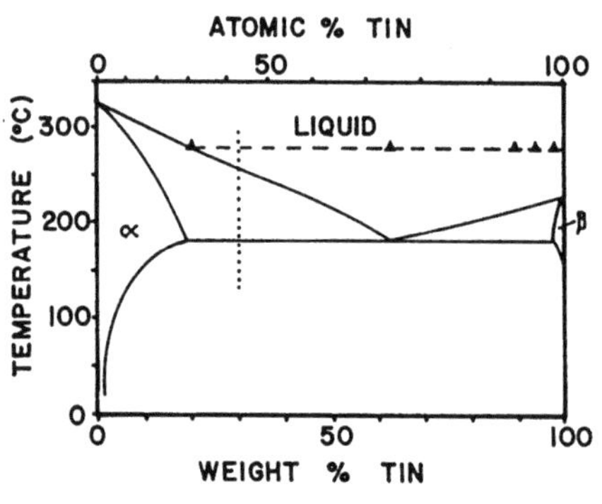

Fig. 1. Bulk phase diagram for a lead-tin alloy (after reference 1)

1.1 Melting in the Bulk

Lead-tin has three phases separated by mixed phase coexistence regions, Fig. 1. The solid phase α consists of a small amount of tin in solid solution in lead whereas solid phase β is lead in solution in tin. At high temperatures, there is a liquid phase in which lead and tin are completely miscible at all concentrations.

+Present address: Fritz-Haber-Institute, Berlin, Germany.

Consider the equilibrium behavior of the bulk alloy as the temperature is varied. The dotted vertical line in Fig. 1 illustrates the case of a 70 w/o Pb alloy. (w/o and a/o stand for weight percent and atomic percent respectively.) Below the eutectic, T_{Eu} = 183°C, the alloy is solid and consists of a mixture of α and β phases as given by the lever rule. At 100°C the α-phase accounts for 72.4 w/o of the alloy and just below T_{Eu} its fraction increases to 86 w/o. The α-phase is also enriched in Sn as T approaches T_{Eu}. However, the average alloy composition remains 70 w/o Pb. Bulk melting is a first order transition which begins at T_{Eu} and is complete when T reaches the liquidus line at T_L = 255°C. During melting the alloy consists of a mixture of α-phase and liquid whose compositions can be calculated using the lever rule. Just above T_{Eu}, the alloy is about 26 w/o liquid and the composition of the liquid is 62 w/o Sn and 38 w/o Pb. As T increases above T_{Eu} the fractional amount of liquid phase increases continuously to a value of 100% at $T = T_L$. Above T_L, the alloy is a single phase liquid.

1.2 Melting at the Surface

The textbook picture of bulk melting given above needs to be modified if only the behavior of the alloy surface is of interest. For example, since the free surface energy of a liquid phase is generally lower than that of the solid, the liquid phase will wet the solid so that any free surface must be

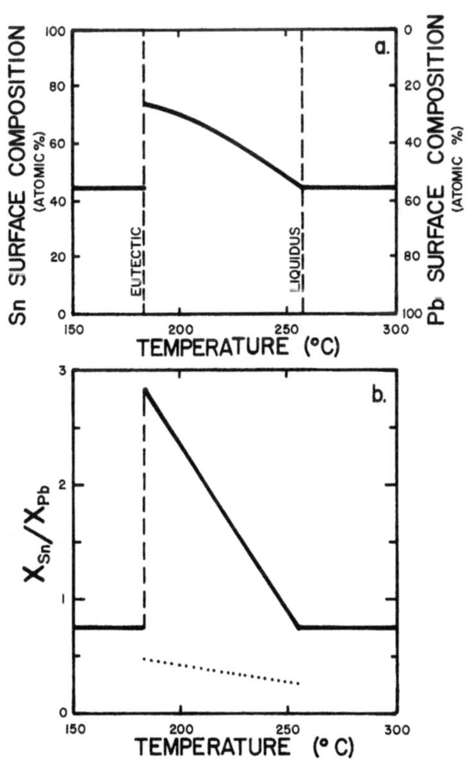

Fig. 2. (a) Surface composition determined from bulk phase diagram, Fig. 1. (b) Ratio of tin-to-lead calculated from (a). The dotted curve is calculated from Fig. 3, see text.

coated with the liquid. In essence, for the case of a 70 $^{w}/_{o}$ Pb alloy, the sample will consist of a pure α-phase solid core which is coated with the liquid phase when $T_{Eu} < T < T_L$.

If the melting is observed with a surface sensitive spectroscopy such as Auger Electron Spectroscopy, the measured composition for $T > T_{Eu}$ will thus be the liquid phase only and <u>not</u> the average composition. Figure 2, illustrates the behavior expected for the 70 $^{w}/_{o}$ Pb case. The Sn composition (in atomic percent) is plotted vs. temperature in the upper panel and the ratio of the Sn and Pb concentrations is shown in the lower panel. The most important feature is the large enhancement of surface Sn concentration in the liquid-solid mixed phase region. Below T_{Eu}, the surface has the average bulk composition if the surface energies of the α and β phases are assumed to be equal. However, as soon as T becomes greater than T_{Eu} the surface is coated with a liquid and the composition jumps to that of the eutectic. As the temperature is further increased toward T_L, the composition of this liquid approaches the bulk composition.

The above discussion has completely ignored surface and grain boundary segregation effects. But it clearly shows that large changes in surface composition should be expected even in their absence.

2. Previous Studies of the Surface Composition of Lead-Tin Alloys

In order to complete the picture of surface melting presented in the previous section it is necessary to include the effects of surface segregation. Therefore, in this section, we briefly review previous studies of the surface properties of the Pb-Sn system. It should be emphasized at the outset that such studies are difficult, primarily because impurities are very mobile at temperatures near the melting point. Thus, even very low bulk impurity concentrations can result in rapid contamination of the sample surface.

The only study of the equilibrium surface composition of liquid Pb-Sn alloys is that of Smith, Seah and Lea [2]. Their results are reproduced in Fig. 3 where the surface Pb concentration X^S_{Pb} is plotted as a function of bulk Pb concentration X^B_{Pb} at a temperature of 275°C. In the absence of surface segregation, the data shown as the vertical bars would lie along the forty-five degree line. Clearly there is very strong segregation of Pb to the surface at all compositions studied. These results are consistent with most theories of surface segregation which predict that components with low surface energy and large atomic volumes will be enriched at the surface. The data points in Fig. 3 have been plotted directly on Fig. 2b neglecting any T-dependence of surface segregation in the liquid. Clearly surface segregation has an extremely strong influence.

Segregation effects are also important for solid alloy surfaces. A 40 $^{w}/_{o}$ (28 $^{a}/_{o}$) Pb alloy was studied by Frankenthal and Siconolfi [3] using Auger Spectroscopy. The clean surface

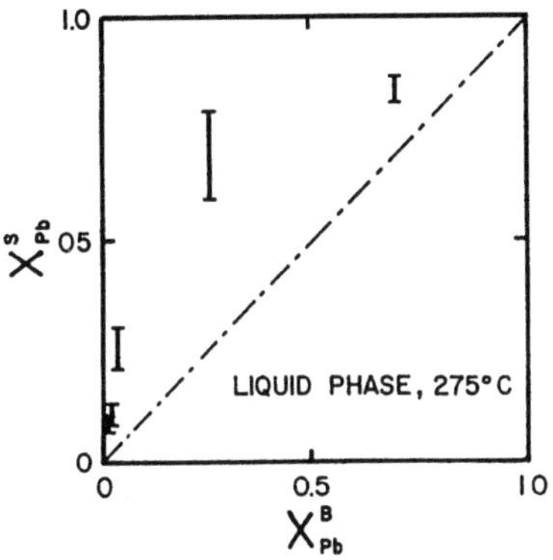

Fig. 3. Surface composition of lead-tin alloys at 275°C (after reference 2).

near room temperature was found to consist of two phases with average enhancement of Pb between 58 $^a/_o$ Pb and 32 $^a/_o$ Pb. More detailed analysis was not possible because the diameter of the electron beam exceeded the mean size of the microstructure of the α and β phases so that the Auger spectrum is an average over α- and β-phases. Times of several hours or longer were required for equilibrium to be reached following sputter cleaning with Argon ions. In order to clarify these results, Frankenthal and Siconolfi [4] also studied pure α- and β-phase alloys. In the case of a 96.92 $^a/_o$ Pb α-phase sample, the equilibrium amount of lead enhancement was found to <u>increase</u> as the temperature was increased over the entire temperature range studied (25°C to 200°C). On the other hand, for a 0.17 $^a/_o$ Pb β- alloy, the relative Pb concentration <u>decreased</u> with increasing temperature.

The effect of sputtering Pb/Sn alloys with 2 keV argon ions has also been reported by Frankenthal and Siconolfi [3,4]. For both α- and β-phase samples they found strong depletion of Pb from the surface region. This depletion was larger at low temperatures for both α-phase and β-phase samples. They interpreted their results as evidence that the higher heat of vaporization of Sn was responsible for its enhancement. Sn enhancement was <u>not</u> observed for any of the samples studied in our work and described below.

Opilla [5] also studied the 96.92 $^a/_o$ Pb alloy. Immediately following sputtering at room temperature, Opilla's analysis of his data yielded a surface composition very near that expected from the bulk sample. Comparing this result with Frankenthal and Siconolfi yields an equilibrium surface composition of about 1.2 $^a/_o$ Sn and 98.8 $^a/_o$ Pb. It should be pointed out that Opilla's surface was strongly contaminated with graphitic carbon which might have significantly altered the surface behavior. Opilla also observed Sn segregation to occur near grain boundaries once sputtering ceased. Segregation continued over hundreds of hours at room temperature. Sn enhancements of up to six-fold were observed. Exposure to about 1200 L O_2 immediately after sputtering increased the rate of Sn segregation near grain boundaries. The possible influence of surface carbon contamination on this segregation was not discussed.

3. Details of the Experiments

The alloys were prepared from 99.999% pure pellets of lead and tin. The amount of each component required to make one lot of 20 g of the alloy was weighed to an accuracy of 0.01 gm or better. The pellets were mixed and melted in an alumina crucible on a hot plate in a glove box purged with flowing argon gas. This significantly reduced the initial amount of oxide contamination on the liquid surface.

The sample was held in a pyrolized graphite bucket 12 mm inner diameter and 2 mm deep. The base and wall of the bucket was 1 mm thick. A shallow groove was cut in the wall around the center of the bucket, and tantulum wires 0.125 mm in dia. encircled the outside of the bucket in this groove. The ends of the wires were spot-welded to two parallel tungsten rods mounted on the end of a transfer rod in order to suspend the bucket between the rods as on a fork. A chromel-alumel thermocouple junction was inserted between the tantulum wire and the graphite bucket on one side.

Before the graphite bucket was mounted on the transfer rod, a sample (3-4 g) of the alloy was melted and poured into it under an argon atmosphere. After the alloy cooled in the bucket, it could be removed as a 2-3 mm thick disk and its surfaces were scraped lightly with a razor blade to remove as much of the surface oxide layer as possible. The disk was then slipped back into the bucket and the bucket mounted on the transfer rod as described above.

Because the clean liquid alloy did not wet the graphite, the shape of the sample went from a flat disk to a more hemispherical ball upon melting in vacuum. In order to prevent the sample from falling out of the bucket after this shape change, one of two schemes was used. In the first case, two tungsten wires crossed the bottom of the bucket and a piece of copper foil about 2 mm x 4 mm x 0.125 mm was attached between them at the center of the bucket. When the sample was melted in the vacuum chamber, the alloy wet the copper and clung tightly, thus being held in place. In the second case, the same tungsten wires were simply crossed over the top of the flat sample disk after it was first refitted into the bucket. After melting in vacuum, the alloy sample remained firmly in place under these wires despite the change in shape.

The sample and its holder were supported on a transfer rod and could be inserted into a separate preparation chamber that was sealed from the main vacuum chamber by a gate valve. The preparation chamber and manipulator for the transfer rod were pumped out and baked separately before this valve was opened. The pressure in the preparation chamber reached about 7×10^{-8} mbar, and the main chamber could be maintained at 2×10^{-9} mbar or lower during the course of the experiments. The sample could be heated by direct current resistive heating of the encircling tantalum wires and cooled by flowing liquid nitrogen through the transfer rod. The temperature range was from about -30°C to above the alloy melting point.

Analysis was done with a E11 Leybold-Hereaus hemispherical analyzer. The analyzer accepted electrons emitted normal to the sample. The the electron gun used for Auger Spectroscopy was mounted at an angle of about 30° from the surface normal of the horizontal sample. The AES scans were taken in the constant retard ratio mode (CRR=4), and each scan from 0-1000 eV kinetic energy took five minutes. The beam energy was 3000 eV with a nominal beam current of 6 μA. Lock-in detection was used with a modulation of 2 V peak-to-peak to obtain the first derivative of the signal versus kinetic energy.

An argon ion sputter gun was attached to the preparation chamber, and the samples could be sputtered in either the solid or the liquid state. In the latter case, the bucket was held horizontally, and the sputter gun was incident at 60° from the surface normal. Typical settings were 2-4 x 10^{-4} mbar argon pressure with 200-300 eV beam voltage to produce 20-30 μA ion current at the sample. Due to its thermal isolation, the sample temperature rose by as much as 120°C during sputtering unless it was cooled with a liquid nitrogen flow in the transfer rod.

4. RESULTS

A typical AES scan for a clean 70/30 wt % Pb/Sn alloy is shown in Fig. 4.

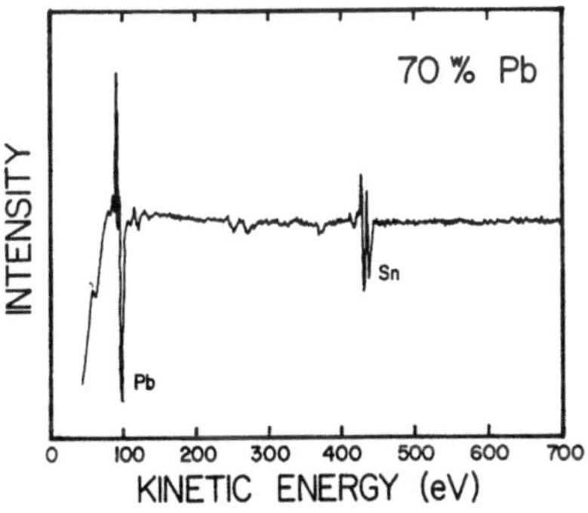

Fig. 4. Auger spectrum of a clean lead (70 w/o) - tin (w/o) alloy

The lead and tin intensity ratios were calculated using the peak-to-peak heights for the Pb peak at 98 eV and for the Sn peak at 434 eV. The two peaks at 250 eV and 275 eV are from lead and are nearly the same size on the clean sample. The presence of any carbon on the surface would be seen in the spectrum by a distinct increase in the size of the peak at 275 eV relative to that at 250 eV. Oxygen would appear at 512 eV. Neither of these contaminates are visible in Fig. 4.

The AES peak-to-peak heights of lead and tin should be directly proportional to their relative atomic concentrations. The average composition in the near surface region sampled by the escaping Auger electrons could be estimated from the measured peak-to-peak heights if the energy dependence of the electron mean-free pathlength and the sensitivity factors for lead and tin were known. Unfortunately, the sensitivity factors could not be precisely determined for our apparatus due to strong changes in the background shape in the AES scan as, for example, the beam position was varied across the sample surface, and due to changes in the shape of the sample surface as discussed in the experimental details section. Within the course of an experiment in which beam and sample position remain fixed, the sensitivity factors remain the same from scan to scan, and the AES peak-to-peak intensity ratio for tin to lead, I_{Sn}/I_{Pb}, can be normalized to a given surface concentration ratio, X_{Sn}/X_{Pb}.

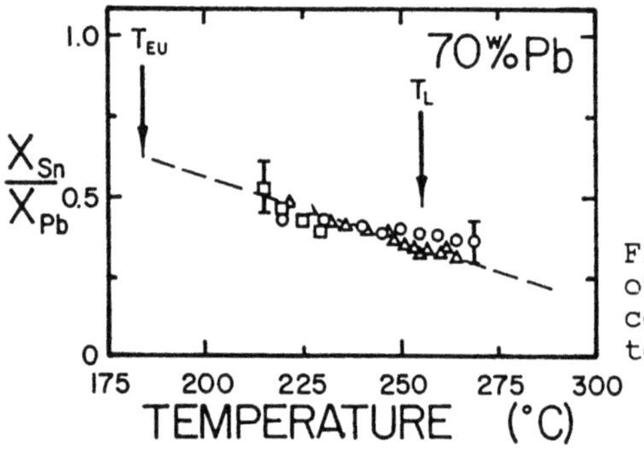

Fig. 5. Variation of the ratio of the surface tin-to-lead compositions near the liquidus temperature.

Three different samples of 70/30 w/o Pb/Sn alloy were subjected to heating or cooling while monitoring the surface composition with AES at various temperatures. The peak-to-peak ratios I_{Sn}/I_{Pb} obtained from each sample are compared in Fig. 5. The data given by circles was obtained while heating a sample from 221°C to 270°C. The temperature was increased in 3-4°C increments and an Auger spectrum was taken after about five minutes at each temperature. The data given by the triangles was obtained while cooling from 270°C to 220°C. At each point between 270°C and 250°C, thirty minutes was allowed for equilibrium to be established. Below 250°C, longer periods of up to 2 hours were used to insure that the ratio I_{Sn}/I_{Pb} reached a nearly constant value. The final set of results, given by the squares, represents a heating experiment in which the sample was taken from 210°C to 230°C. Approximately thirty minutes was allowed at each temperature. The three data sets were normalized to have the same average value of I_{Sn}/I_{Pb} in the temperature range of overlap, 220°C to 230°C. The measured I_{Sn}/I_{Pb} were then converted to X_{Sn}/X_{Pb} by normalizing at 225°C to the dotted curve in Fig. 2b. This curve was derived from the results of Smith, Seah and Lea [2] as described earlier. Clearly the functional form is identical, thus strongly supporting the picture of surface melting presented in Section 2. No contaminants were present on the surface for the data of Fig. 5

In only one case was the surface composition measured during heating and cooling through the eutectic point, 183°C. The data is shown in Fig. 6 where the ratio I_{Sn}/I_{Pb} is plotted

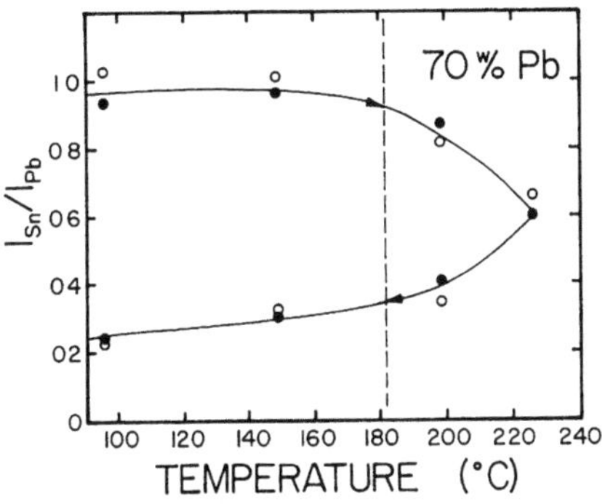

Fig. 6. Variation of the ratio of tin-to-lead Auger signals near the eutectic temperature (vertical dashed line) versus temperature for data obtained five minutes (open dots) and thirty minutes (solid dots) after each temperature was reached. The arrows show the direction of temperature change. Clearly, thermodynamic equilibrium was not achieved over the entire temperature cycle. However, the measurements already shown in Fig. 5, indicate that surface equilibrium is reached within thirty minutes for T above about 220°C. Therefore, the data point at 225°C in Fig. 6 should also represent equilibrium conditions. Although a similar conclusion about equilibrium cannot be made for the other data points in Fig. 6, it is clear that the Sn surface composition is reduced upon cooling below 220°C. Surprisingly, this decrease occurs at a temperature above the eutectic! Visually, the samples appeared to develop a liquid layer at least 5-10°C below the liquidus temperature.

Isothermal annealing experiments were carried out to study the effects of sample pretreatment on the kinetics of ordering and to determine the time required for the surface composition to reach equilibrium.

Figure 7 shows the results of two measurements which illustrate the effects of sample pretreatment. In the case of curve (a), the sample was first sputter cleaned at a temperature below room temperature and then immediately heated to 197°C. Curve (b) was obtained under the same conditions as (a) except the sample was maintained near room temperature for three hours prior to the isothermal anneal. Two important features are apparent in Fig. 7. First, some segregation of Sn to the surface occurs during the three hour room temperature anneal. Second, the same steady state composition is reached at 197°C, even though the detailed surface compositions change in very different ways during the approach to equilibrium. About 80-100 minutes are required to reach equilibrium at this temperature.

Similar isothermal measurements were carried out for many temperatures. In general the time to reach equilibrium was found to decrease with increasing temperature and the average surface Pb concentration increased at higher T in agreement with the earlier studies on a 96.92 a/o Pb alloy [4].

Data such as that in Fig. 7 also provide information about preferential sputtering. In this case we find the average

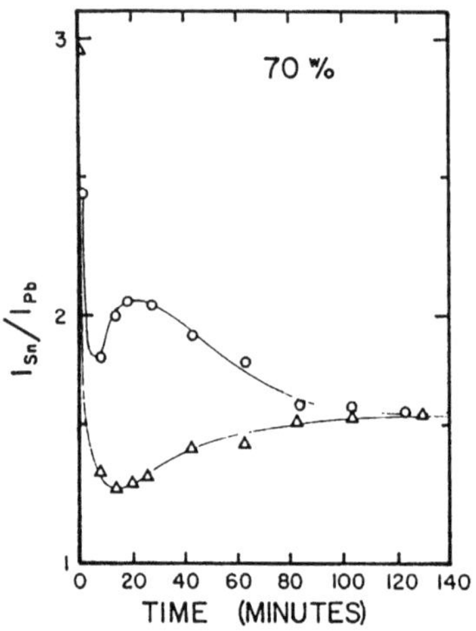

Fig. 7. Effects of pretreatment on the approach to equilibrium. See text for details.

surface Pb concentration begins to decrease immediately after sputtering is ceased. Therefore, we conclude that Sn must be preferentially sputtered from the surface. Our other experiments lead to the same conclusion.

5. Discussion

The simple theory outlined in Sections 1 and 2 for the surface composition of a lead-tin binary alloy during melting predicts specific behavior. Below either eutectic or solidus temperatures, the alloy is solid and the surface concentration is determined by surface segregation processes in the α- and β-phase microdomains. As shown by Opilla [5], surface diffusion of Sn from grain boundaries competes against the preferential segregation of Pb from the bulk phase. The time required for these processes to reach equilibrium is very long, perhaps hundreds of hours. Thus, it is unlikely that free surfaces are ever at equilibrium in real alloys.

Once the melting temperature (T_{Eu}) is exceeded, the equilibrium configuration consists of pure α-phase solid coated with a Sn-rich liquid for alloys with Sn concentration less than the eutectic concentration. The experiments described above for a 70 $^{w}/_{o}$ Pb alloy provide compelling evidence for this picture, at least within 5-10°C of T_L, and are in good agreement with previous studies of the liquid phase by Smith, et al. [2]. However, as illustrated in Fig. 8 and by the visually observed changes in the surface, the liquid layer does not seem to form until 20-30°C above the eutectic.

This raises the question of the nature of surface segregation at a solid-liquid-vacuum interface. In general, diffusion in the liquid should be fast compared to diffusion in the solid phase. Therefore, the ability of the liquid phase and its surface to reach equilibrium will be determined by diffusion in the solid phase and by segregation phenomena at the interface between the solid and its liquid coating. Thus, the thirty minutes allowed for equilibration in Fig. 6 may have been insufficient for true equilibrium to be achieved. This question can only be resolved by further research.

Acknowledgement

This work was supported by the Laboratory for Surface Science and Technology and by a grant from IBM-Endicott.

References

1. Metals Reference Book, Eds. C. J. Smithells and E. A. Brandes (Butterworths, Boston, 1976).
2. G. C. Smith, M. P. Seah and C. Lea (to be published).
3. R. P. Frankenthal and D. J. Siconolfi, J. Vac. Sci. Technol. 17, 1315 (1980).
4. R. P. Frankenthal and D. J. Siconolfi, Surface Sci. 104, 205 (1981).
5. R. L. Opilla, J. Vac. Sci. Technol. A4, 173 (1986).

Part III

Theoretical Models for Segregation and Diffusion

Recent Developments in the Theory of Many-Body Processes: Implications for the Study of Surface Dynamics

J.D. Doll[1], *D.L. Freeman*[2], *and A.F. Voter*[3]

[1]Los Alamos National Laboratory, Mail Stop G-738, Los Alamos, NM 87545, USA
[2]Department of Chemisty, University of Rhode Island, Kingston, RI 02881, USA
[3]Los Alamos National Laboratory, Mail Stop J-569, Los Alamos, NM 87545, USA

1 Introduction

As the readers of these words are undoubtedly aware, issues of interfacial structure and dynamics are central to a wide range of phenomena having both conceptual and practical significance. As a consequence, the study of such phenomena is important, both from the viewpoint of fundamental theory and from the viewpoint of more practical applications.

Prompted in part by the development and refinement of a variety of new experimental probes, we have witnessed a substantial increase in activity in surface science over recent years. Of particular interest for the present discussion has been the development of methods capable of probing the diffusion and aggregation of surface adsorbates. Many of these experimental methods are discussed in detail elsewhere [1-4].

In the present work we would like to draw attention to a series of recent theoretical developments that have important implications for the study of interfacial processes. Some of these developments have their origins within surface science, but many have taken place in other arenas, and are thus likely to be somewhat unfamiliar to the reader. In what follows we present brief sketches of some of these theoretical developments, aiming the discussion primarily at the non-specialist. It is our hope that such introductory, overview discussions will promote more widespread appreciation and application of these methods. More complete discussions of the following methods as well as examples of their application are available elsewhere [5-9].

Section 2 considers the particular problem of adsorbate diffusion. Significant developments in "rare event dynamics" have made it possible to formulate the surface diffusion problem in a formally exact and computationally convenient fashion. Important in its own right, diffusion is also the prototype for more general activated processes such as desorption and reaction. Section 3 explores a variety of recent developments in quantum-mechanical path integral theory. Loosely speaking, these methods are the quantum analogs of the familiar classical Monte Carlo/molecular dynamics techniques and provide a general and practical route from microscopic interactions to macroscopic observables. The unique ability of these path-integral methods to deal with many-body quantum-mechanical systems provides a new and important tool in the analysis of condensed matter problems in general and interfacial problems in particular.

In what follows we concentrate principally on dynamical methods. The logical separation of the dynamics from the construction of the microscopic force laws is a useful one since the execution of the dynamics is largely independent of the detailed nature of the interaction. Many aspects of the important problem of surface electronic structure are reviewed elsewhere [10].

2 Rare-Event Dynamics

For many problems in interfacial dynamics classical mechanics provides a suitable bridge linking microscopic force laws and macroscopic observables. In such situations well-developed equilibrium Monte Carlo [11] and time-dependent molecular dynamics methods [12] provide general and quite powerful tools for analysis. In principle, once the microscopic force laws are specified such problems are "solved". In practice, however, this formal program is often difficult to realize.

Activated processes represent a general class of problems that frustrate a direct application of these otherwise well-developed classical methods. The essence of the difficulty in such problems is the inherent disparity of time scales present in the dynamics. To properly describe the event of interest (e.g. a diffusive hop of an adsorbate, a bond rupture, a reactive event), it is necessary to resolve the dynamics on a time scale that is short compared with the relevant molecular motions. Time steps of the order of 10^{-15} sec are typical for molecular dynamics applications. With such time steps present computer performance limits the total time over which molecular dynamics trajectories can be followed to something on the order of 10^{-9} sec or less. If the interval between events of interest is larger than 10^{-9} sec, a common situation for activated processes, then direct molecular dynamics studies are not generally practical.

As is described in detail elsewhere, the "rare-event" problem has been solved for many applications of interest in interfacial dynamics [6]. We draw special attention here to the important example of adsorbate diffusion. This problem of diffusion in solids was first approached using rare-event methods by Bennett [13]. The methods described below were formulated by Voter, et al. [6] and draw heavily from the work of Chandler and co-workers [14,15]. Jacucci and co-workers have independently pursued similar lines of developments [9].

The key element in formulating a theory of rare-event dynamics is to organize the problem in such a way that only short-time dynamics are explicitly required. In the case of diffusion this seemingly contradictory objective is accomplished by decomposing the diffusion coefficient as a transition state theory (TST) estimate [16-20] times a correction factor,

$$D = D_{TST} {}^{*}(D/D_{TST}) \qquad (1)$$

This decomposition accomplishes several related objectives. Most importantly, it places the bulk of the burden of computing the diffusion coefficient on TST, an <u>equilibrium</u> method. Monte Carlo techniques make possible the general numerical solution of such equilibrium problems and, in particular, permit the calculation of exact TST surface diffusion coefficients [21,22]. With TST performing the bulk of the labor, the dynamical demands of the method are reduced to those that are necessary to remove the errors in TST. The central point is that these TST errors, errors that arise from that theory's approximate treatment of recrossing and correlated, multiple hopping events, can be quantified <u>using only short-time dynamics</u>.

It is important to emphasize that the above approach produces numerically exact classical results for the particular microscopic force law utilized. Furthermore, the theory places no special restrictions on the functional form of the force law.

The approach outlined above has been used to compute single adsorbate self-diffusion coefficients [5,6]. Figure (1) illustrates a particular example of such an application. Shown there is the self-diffusion coefficient computed for the Rh/Rh(100) system using a particular model of the Rh-Rh interactions. The calculations presented in Fig. 1 cover temperature regimes where direct molecular dynamics simulations would have been quite impractical.

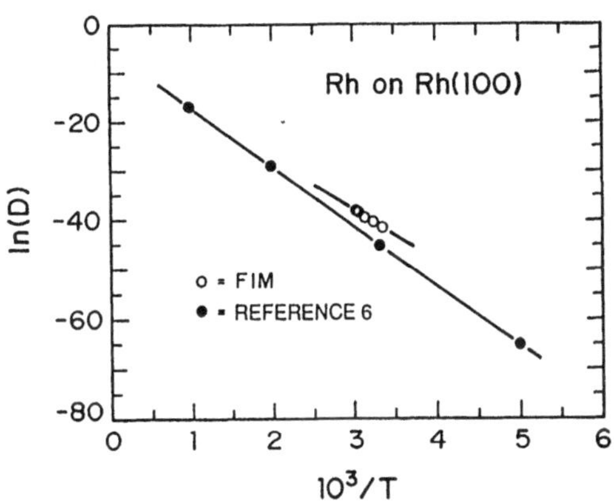

Fig. 1. Surface self-diffusion coefficients for Rh on Rh(100), plotted in Arrhenius form. The units are D [cm^2/s] and T [K]. The error bars (two standard deviations) are not shown since they are smaller than the plotted points. Experimental results (open circles) are from G. Ayrault and G. Ehrlich, J. Chem. Phys. 60, 281 (1974) and theoretical results are from Ref. (6).

As emphasized above, the numerical results of Fig. (1) correspond to exact classical-mechanical results for the particular potential model utilized. Confident that the dynamics have been properly treated, a comparison of the experimental and computed results in Fig. (1) provides unambiguous insight into the shortcomings of the approximate potential model utilized. The rigorous implementation of the dynamics possible with the present generation of numerical methods opens a new dimension in the analysis of activated processes. We refer the interested reader to Refs. [5,6] for more complete details.

Voter has recently utilized methods related to those described above to develop a general theory of overlayer dynamics [23] (i.e. dynamics of multiple, interacting adsorbates). This theory has the important feature that it permits the study of adsorbate dynamics over macroscopic time scales. Unlike many phenomenological statistical methods, however, the dynamical quantities that enter the theory are derived from the fundamental microscopic interactions.

3 Monte-Carlo Path Integral Methods

The classical-mechanical methods described in the previous section are appropriate for many problems in interfacial dynamics. For others, however, the quantum-mechanical effects are sufficiently large that by themselves classical methods are inadequate. The dynamics of small mass adsorbates and of quantized vibrations are examples of this latter class of problems.

Substantial progress has been made recently in the treatment of many-body condensed-phase quantum mechanical problems using numerical path integral methods [8,24]. These methods hold the promise of providing general and quite powerful tools for the analysis of both equilibrium and dynamics problems of immediate interest to the interfacial community. Although there are indications that these methods might also be of utility in the treatment of problems in electronic structure [25], the present discussion will presuppose the independent specification of interatomic and intermolecular force laws.

We consider first the application of path-integral methods to equilibrium problems. A typical approach to equilibrium problems is to use numerical Monte Carlo methods to estimate relevant statistical-mechanical thermodynamic averages. Schematically, if the average of interest is given by

$$\langle f \rangle = \frac{\int dx\, \rho(x)\, f(x)}{\int dx\, \rho(x)}\,, \tag{2}$$

one estimates <f> as

$$\langle f\rangle = \frac{1}{N}\sum_{n=1}^{N} f(x_n)\ . \tag{3}$$

In Eq. (2) $\rho(x)$ is the appropriate equilibrium distribution. The Monte Carlo quadrature points, x_n, are generated by means of well-established statistical methods, methods designed to produce points whose probability of occurrence is proportional to the statistical weight, $\rho(x)$. For the canonical ensemble, $\rho(x)$ is given (to within irrelevant constants) classically (CM) and quantum-mechanically (QM, distinguishable particle limit) as

$$\rho(x) = \begin{cases} e^{-\beta V(x)}, & \text{CM} \\ \langle x|\, e^{-\beta H}|x\rangle, & \text{QM} \end{cases} \tag{4}$$

where $V(x)$ is the potential energy function, H is the system Hamiltonian and β is the reciprocal temperature ($1/k_BT$). Equation (2) is written here in a pseudo one-dimensional notation for simplicity only. The essential strength of Monte Carlo methods is their ability to cope with large dimensional problems of precisely the type in Eq. (2).

The availability of the classical distribution, $\rho(x) = \exp(-\beta V(x))$, makes the application of Monte Carlo methods to the classical equilibrium problem quite direct. By contrast, the quantum-mechanical statistical distribution is not generally available analytically. Path integral methods [26] demonstrate, however, that the quantum-mechanical distribution function can itself be written as as an average,

$$\rho(x) = \frac{\int da\ \rho(x,a)\ e^{-\beta\langle V\rangle}}{\int da\ \rho(x,a)}\ . \tag{5}$$

Explicit expressions for the various quantities in Eq. (5) are given elsewhere. Equation (5) is the quantum-mechanical generalization of the simple, classical Boltzmann factor. The averages in in Eq. (5) are over "auxiliary" degrees of freedom that properly account for the quantum-mechanical dispersion about the classical position, x. These auxiliary degrees of freedom describe "paths" that begin and end at the classical position, x. The statistical significance of each path is measured by $\exp(-\beta\langle V\rangle)$, where $\langle V\rangle$ is the average of the potential energy evaluated along that particular path. The classical limit of $\rho(x)$ is recovered by noting that if the dispersion of the auxiliary degrees of freedom about the point x tends to zero, then $\langle V\rangle = V(x)$.

The important point for the present discussion is that the path integral methods render the equilibrium quantum-mechanical problem soluble. The quantum-mechanical problem in the variable x is turned into what is essentially a classical problem in an enlarged space [27], a space that includes not only x, but auxiliary degrees of freedom as well. Specifically, if we combine Eq. (5) and Eq. (2), then the quantum-mechanical equilibrium average of f takes on the essentially classical form,

$$\langle f\rangle = \frac{\int dx\, da\, \rho(x,a)\, e^{-\beta\langle V\rangle} f(x)}{\int dx\, da\, \rho(x,a)\, e^{-\beta\langle V\rangle}} . \tag{6}$$

Monte Carlo methods provide a general method for evaluating averages of the type in Eq. (6) and hence provide a general means for treating equilibrium quantum-mechanical problems.

Numerous applications of quantum Monte Carlo methods are discussed elsewhere [8,24]. As emphasized by Abraham and Broughton [28], an important area of application of equilibrium path integral methods to interfacial problems is to phase equilibria. Another class of interfacial problems that is an obvious candidate for study using such methods is heterogeneous nucleation, where generalizations of published work on homogeneous nucleation [29] are possible.

Although the basic line of development within the path integral community is toward numerically exact methods, we mention here one particularly appealing semiclassical result. As shown by Feynman and Hibbs [30], the semiclassical limit of $\rho(x)$ is given by

$$\rho(x) = e^{-\beta V_{eff}(x)}, \tag{7}$$

where

$$V_{eff}(x) = \frac{\displaystyle\int_{-\infty}^{\infty} dy\, e^{-y^2/2\sigma^2} V(x+y)}{\displaystyle\int_{-\infty}^{\infty} dy\, e^{-y^2/2\sigma^2}}, \tag{8}$$

with

$$\sigma^2 = \frac{\hbar^2}{12mk_BT}. \tag{9}$$

In this classically structured expression (c.f. Eq. (4)) quantum effects enter through the effective potential, $V_{eff}(x)$. This effective potential accounts in an approximate way for the quantum-mechanical dispersion of the particle by replacing the value of the potential energy at the classical position, V(x), by its average over a "neighborhood", a neighborhood whose dimension is dictated by the de Broglie wavelength. Once this mass and temperature-dependent effective potential has been constructed, ordinary classical algorithms can be employed to extract observables. The connection of this effective potential approach with more familiar semiclassical methods is discussed elsewhere [31].

A number of applications of the semiclassical effective potential formalism to surface dynamics have been reported. Such methods have been utilized to compute surface diffusion coefficients for hydrogen and its isotopes on a variety of metal surfaces [32-34]. The results of such diffusion applications have been surprisingly successful, both in terms of correlating with experimental data and in terms of quantitative agreement with the results of other theoretical methods for model calculations. The degree of success of such applications suggests that the effective potential approach offers a useful intermediate level of theory, bridging purely classical and purely quantum-mechanical approaches.

Methods related in spirit to the effective potential methods described above have recently been presented by Gillan [35,36]. These developments represent an important stride towards a complete quantum-mechanical theory of hydrogen diffusion in solids.

In closing, we draw attention to important recent developments in the field of quantum dynamics. For some time it has been appreciated that the path integral formulation of quantum mechanics sketched above for equilibrium problems can be extended to treat problems in quantum dynamics. This extension is formally accomplished by making the parameter β complex, associating the real and imaginary parts of β with temperature and time, respectively. The practical difficulty with such an approach has historically been that the numerical Monte Carlo methods, successful in the equilibrium context, are overwhelmed in the dynamics problem. The practical difficulty centers on the fact that the analogs of the classical Boltzmann factors in Eq. (6) are oscillatory whenever β is complex. Since the imaginary part of β is proportional to the physical time in the problem, these oscillations become more serious for larger real times, effectively limiting the times over which such problems can be studied using ordinary Monte Carlo methods to

$$t < \beta\hbar. \tag{10}$$

Generalizations of ordinary Monte Carlo methods have recently been reported that appear to solve the essential numerical difficulties associated with the rapid oscillations in the path integral formulation of the quantum dynamics problems [37-42]. Preliminary applications of these methods to model problems have been completed. The results of these applications indicate that there is reason for optimism concerning the scope and utility of the new approaches. Although more work will be required before these new Monte Carlo methods are fully characterized, these new methods offer the tantalizing possibility of making practical the general, numerically exact study of many-body quantum dynamics. If this current hope is realized, several classes of important dynamics problems, including many of interest to the interfacial community, will be opened for study.

References

1. G. Ehrlich, CRC Crit. Rev. Solid State Mater. Sci. **10**, 391 (1982).
2. G. Ehrlich and K. Stolt, Ann. Rev. Phys. Chem. **31**, 603 (1980).
3. J. A. Panitz, J. Phys. E, **15**, 1281 (1982).
4. S. M. George, M. DeSantolo and R. B. Hall, Surf. Sci. **159**, L425 (1985).

5. J. D. Doll and A. F. Voter, Ann. Rev. Phys. Chem. **38**, 413 (1987).
6. A. F. Voter and J. D. Doll, J. Chem. Phys. **82**, 80 (1985).
7. M. Toller, G. Jacucci, G. DeLorenzi and C. P. Flynn, Phys. Rev. **B32**, 2082 (1985).
8. B. Berne and D. Thirumalai, Ann. Rev. Phys. Chem. **37**, 401 (1986).
9. J. D. Doll and D. L. Freeman, Science **234**, 1356 (1986).
10. A. M. Stoneham and J. H. Harding, Ann. Rev. Phys. Chem. **37**, 53 (1986).
11. J. P. Valleau and S. G. Whittington, in Modern Theoretical Chemistry, B. J. Berne, ed (Plenum, New York, 1977), Vol. 5, pp 137-168.
12. J. Kushick and B. J. Berne, in Modern Theoretical Chemistry, B. J. Berne, ed. (Plenum, New York, 1977), Vol. 6, pp 41-63.
13. C. H. Bennett, in Diffusion in Solids, A. S. Nowick, ed. (Academic, New York, 1975), p 73.
14. D. Chandler, J. Chem. Phys. **68**, 2959 (1978).
15. J. A. Montgomery, D. Chandler and B. J. Berne, J. Chem. Phys. **70**, 4056 (1979).
16. S. Glasstone, K. Laidler and H. Eyring, The Theory of Rate Processes (McGraw-Hill, New York, 1941).
17. G. Vinyard, J. Phys. Chem. Solids **3**, 121 (1957).
18. M. Kac, Am. J. Math. **65**, 609 (1943).
19. N. B. Slater, Theory of Unimolecular Reactions (Cornell University Press, Ithaca, 1959).
20. J. D. Doll, J. Chem. Phys. **73**, 2760 (1980).
21. A. F. Voter and J. D. Doll, J. Chem. Phys. **80**, 5814 (1984); ibid, 5832 (1984).
22. A. F. Voter, J. Chem. Phys. **82**, 1840 (1985).
23. A. F. Voter, Phys. Rev. **B34**, 6819 (1986).
24. J. Stat. Phys. **43**, 729-1244 (1986).
25. D. Ceperley and B. Alder, Science **231**, 555 (1986).
26. R. P. Feynman and A. R. Hibbs, Quantum Mechanics and Path Integrals (McGraw-Hill, New York, 1965)
27. D. Chandler and P. Wolynes, J. Chem. Phys. **74**, 4078 (1980).
28. F. Abraham and J. Broughton, Phys. Rev. Lett. **59**, 64 (1987).
29. D. L. Freeman and J. D. Doll, J. Chem. Phys. **82**, 462 (1985).
30. Ref. [25], Sect. (10.3).
31. J. D. Doll and L. E. Myers, J. Chem. Phys. **71**, 2880 (1979).
32. S. Valone, A. F. Voter and J. D. Doll, Surf. Sci. **155**, 687 (1985).
33. S. Valone, A. F. Voter and J. D. Doll, J. Chem. Phys. **85**, 7480 (1986).

34. C. Mak and S. M. George, Chem. Phys. Letts. **135**, 381 (1987).
35. M. J. Gillan, Phys. Rev. Lett. **58**, 563 (1987).
36. M. J. Gillan, J. Phys. C: Solid State Phys. **20**, 3621 (1987).
37. J. D. Doll and D. L. Freeman, Adv. Chem. Phys. (in press).
38. J. D. Doll, R. D. Coalson and D. L. Freeman, J. Chem. Phys. **87**, 1641 (1987).
39. J. D. Doll, D. L. Freeman and M. J. Gillan, Chem. Phys. Letts. (in press).
40. V. S. Filinov, Nucl. Phys. **B271**, 717 (1986).
41. J. Chang and W. H. Miller, J. Chem. Phys. **87**, 1648 (1987).
42. N. Makri and W. H. Miller, Chem. Phys. Letts. **139**, 10 (1987).

Computer Simulation of Segregation and Diffusion at Oxide Interfaces

P.W. Tasker[1] *and W.C. Mackrodt*[2]

[1]Theoretical Physics Division, Harwell Laboratory, Oxfordshire OX11 0RA, England
[2]ICI plc, P.O.Box 11, The Heath, Runcorn, Cheshire WA7 4QE, England

Abstract Static lattice calculations of ionic materials can give quantitative prediction and interpretation of experiment. The method is briefly reviewed in this paper and two examples of recent applications to interface properties are given. Firstly, we consider impurity segregation to the crystal surface. Calculations of isovalent impurities (particularly calcium) in magnesium oxide and aliovalent magnesium in α–alumina are discussed and compared with recent experimental data. Questions of the defect vibrational entropy and the consequences of defect charge on surface-defect interactions is considered. Secondly, we discuss calculations on diffusion down grain boundaries in nickel oxide. These studies are aimed at interpreting the apparent enhanced diffusion observed in polycrystalline scales growing on the oxidising metal. The grain boundary diffusion is thought to determine the oxidation rate.

1. Introduction

Understanding the behaviour of surfaces and interfaces in ionic materials is the key to interpreting many important properties. Ceramics are often largely ionic and grain boundaries play a predominant role in determining the properties of these polycrystalline materials. Greater understanding is needed in areas such as processing (sintering), mechanical properties (fracture, toughness), catalysis and catalyst supports, adhesion, oxidation and film growth etc. However despite its importance, the recent growth in surface science and the application of new surface analytical techniques have still left ionic materials relatively neglected. This is, in part, due to the experimental difficulties encountered in working with insulating materials although some progress has been made /1–3/. Also, many of the real properties of interest are determined by the ever present and poorly characterised impurity concentrations in these materials. From a theoretical viewpoint, ionic materials have a number of advantages. The atomic interactions are better known than for many materials and the experience from bulk defect studies gives us confidence in our ability to predict properties quantitatively /4/. However, although the dominant Coulomb interaction is a straightforward pair potential, its long range creates computational problems. This accounts for the importance of static lattice calculations in this field and the relatively small impact made so far by molecular dynamics. In this paper, we shall describe the application of static calculations to surface and interface studies of oxide materials. To illustrate the sort of properties that can now be studied, we discuss impurity segregation to crystal surfaces and diffusion along grain boundaries.

2. Static Lattice Calculations

A static calculation starts with a model of the crystalline material in which each atom interacts with the others via an interatomic potential. A perfect crystal should be in equilibrium with no forces acting on any atom. When a defect, such as a point defect (e.g. a vacancy or an impurity) or an extended defect (e.g. a surface or grain boundary) is introduced this equilibrium is disturbed. The calculation proceeds by searching for a new atomic configuration in which all the atoms are again in equilibrium. At the end of the relaxation procedure, the structure and energy of the crystal are determined. The calculation ignores thermal effects, but the internal energy at low temperature is often a good approximation to the defect enthalpy /5/. Temperature effects can be included by calculating separately the lattice dynamics of the system and allowing for lattice expansion in the quasi-harmonic approximation. Knowledge of the internal energy and defect entropy enables comparison to be made with higher temperature defect free energies. A recent comparison between a static calculation with lattice dynamics and a molecular dynamics simulation that calculates the free energy more rigorously produced excellent agreement in a study of the oxygen defects in cobalt oxide /6/.

Static calculations have proved to be one of the most valuable methods for estimating segregation behaviour at interfaces in ionic materials. In comparison with other types of material, e.g. metals or semiconductors, the interatomic potentials are better known and provide a more reliable description of the material. This is due, in part, to the true pair-like nature of the Coulomb potential which is the leading term in the interatomic interaction. However, the long-range nature of the interaction has made the application of molecular dynamics to ionic systems of some complexity particularly difficult. Even in a static calculation, special techniques have to be used to handle the long-range sums. Since the component atoms of an ionic material are charged, point defects such as vacancies and impurities are also charged, in general. This produces a long-range polarization of the surrounding crystal that cannot be simulated explicitly. A generalization of the Mott–Littleton approximation /7/ is used and this is illustrated in Figure 1. This shows schematically how a calculation of a point defect near an ionic grain boundary would be carried out. The grain boundary structure is calculated first using the two dimensional periodicity of the system. Then the point defect (e.g. the impurity) is introduced onto a site near the interface. A spherical region of crystal around the defect is modelled explicitly with every ion relaxed to equilibrium. Beyond that, the crystal is treated as a dielectric continuum which is polarized by the defect charge. The size of the inner explicit region can be increased until the defect energy does not vary. Typically, an inner region containing less than 150 ions is sufficient.

The reliability of the final result depends on the realism of the interatomic potentials. These consist of three parts.

a) The Coulomb interaction between charged ions of charges Q_1 and Q_2 is given by,

$$V(r)=\frac{Q_1Q_2}{r}$$

b) A short-range interaction that contains a repulsive component at close distance is required for crystal stability. A common analytical form is

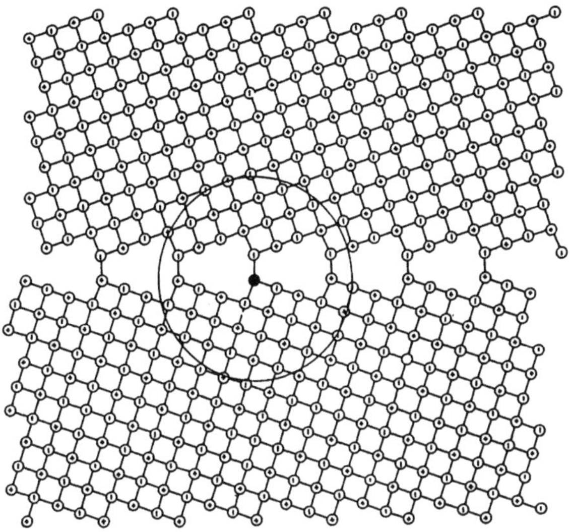

Figure 1 Schematic diagram of the relaxation region around a defect ion (marked as a solid circle) near a grain boundary in an ionic crystal. Within a spherical region, the ions are explicitly relaxed to equilibrium whilst outside this region a continuum approach is used.

$$V(r) = A\exp(-r/\rho) - C/r^6.$$

An analytical form is not essential and numerical potentials are often used.

c) The polarizability of each ion must be included if the dielectric properties of the material are to be reproduced This is usually incorporated via the shell model in which each ion is represented by a core and shell of opposite charge and coupled by an harmonic potential. Displacements of the shell relative to the core lead to the development of a dipole moment on the ion sites. The short-range potential acts between shells and thus couples the interatomic and intra-atomic interactions in a physically sensible way. A good description of the lattice polarization induced by charged defects is essential to reliable defect calculations in an ionic material.

The potentials for a particular material are obtained in either by fitting empirically to the parameters (e.g the A, ρ, C and k) to the known perfect lattice properties /8/ or by calculating them by some quantum mechanical procedure /9/. The latter method is of increasing popularity as the simulation methods are applied to poorly characterised materials. It also provides a consistent route to the treatment of impurity interactions in a host material. In either case, we should obtain good agreement with the bulk structure, energy, elastic and dielectric constants and bulk vibrational phonons. The potentials are not modified in defect or interface calculations so the method is entirely predictive. In fact, interatomic potentials might be expected to vary somewhat at the surface (for example) and this effect is neglected in order to achieve predictions. Such changes should be less in an ionic material than in either a metal or semiconductor and this seems to be born out by the quality of results obtained for surface energies, surface phonons, surface structure and segregation energies in those cases where good experimental data exist.

3 Impurity Segregation

3.1 Segregation Isotherms

Before we discuss the application of lattice calculations to the prediction of impurity concentrations near a crystal surface, we must relate the energies that we can calculate to the experimentally observed quantities. This is done through a statistical mechanical model of the impure system. For a system with a single impurity on one sublattice of the crystal (for example, calcium substituting for magnesium in magnesium oxide, MgO) we can consider the crystal as a two component system and ignore the oxygen sublattice. The free energy can be written as,

$$G=n_1^b g_1^b+n_1^s g_1^s+n_2^b g_2^b+n_2^s g_2^s-kT\ln\Omega$$

in which $n_i^{b,s}$ are the number of bulk and surface ions of type i with individual free energies $g_i^{b,s}$. The last term is the configurational entropy usually calculated with a random distribution over surface and bulk sites. Minimising the free energy with respect to the surface concentrations yields the following expression.

$$\frac{n_1^s}{n_2^s}=\frac{n_1^b}{n_2^b}\exp(-\Delta g/kT)$$

which is the usual Langmuir isotherm. Δg is given by

$$\Delta g=(g_1^s-g_2^s)-(g_1^b-g_2^b)$$

If we neglect the vibrational entropy of segregation, Δg is just the difference between the surface impurity substitutional energy and the corresponding energy in the bulk /10/.

This approach contains a number of assumptions including only one surface site that is different from the bulk and a neglect of impurity-impurity interactions. We can easily see that the last point is rarely true since it predicts that Δg should be independent of defect coverage. Figure 2 shows the calculated heats of segregation, Δh as a function of coverage for calcium impurity in magnesium oxide. It is clearly not constant. If we follow the same simple model, but allow for a variation of Δg with coverage, we obtain a modified Langmuir isotherm /11/.

$$\frac{n_1^s}{n_2^s}=\frac{n_1^b}{n_2^b}\exp(-[\Delta g+n_1^s d\Delta g/dn_1^s]/kT)$$

We can use this expression to calculate the defect concentrations at the surface and over a limited temperature region we may be able to draw a straight line through the Arrhenius plot. This is analogous to the treatment of experimental data and a direct

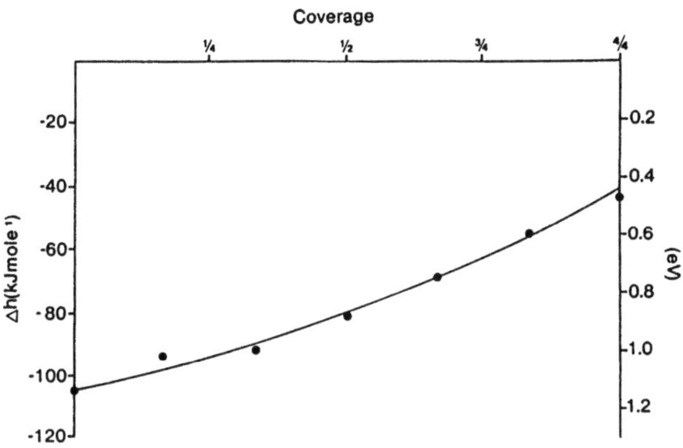

Figure 2 Calculated energies of segregation for a calcium impurity concentrating at the (001) surface of magnesium oxide.

comparison is possible. However, the heat of segregation deduced from the slope of the line is not simply the difference in surface and bulk defect energies that we have calculated. This emphasises the importance of calculating the surface concentrations for direct comparison with experiment.

Finally, we should note that the analysis given here applies to uncharged impurities, for example isovalent substitutional ions. When the impurity is charged, the primary interactions are Coulombic and long-range. Similar interactions occur between the intrinsic defects in the crystal and they give rise to a surface charge and a compensating space-charge region. Treatment of this problem is more complex and will not be discussed in detail here. We note, however, that the binding energy between oppositely charged defects is greatly enhanced in the surface region and so even charged impurities can often be treated as neutral clusters.

3.2 Isovalent Impurities

The segregation of several isovalent impurities to the surfaces of ionic materials have recently been studied using computer calculations /11,12/. The expected trends of large ions segregating to the surface while small ions are retained in the crystal bulk is calculated for simple systems such as impurities in MgO. However, this often assumed behaviour is not predicted in crystals such as alumina or chromia where there is substantial relaxation of the pure surfaces /13/. In these cases, the interaction between the surface relaxation and the impurity segregation can give rise to more complex segregation behaviour.

Our first example is one the most simple ionic systems and yet illustrates some of the problems associated with predicting segregation behaviour. Calcium as a segregating impurity in magnesium oxide is one of the few systems that has been studied quantitatively by experiment. McCune and Wynblatt have used both Auger spectroscopy and Low Energy Ion Scattering to determine the segregation isotherm /1,2/. It is also a fairly straightforward system to calculate and MgO shows little surface relaxation to complicate the segregation behaviour /14/. The difference between the energy to substitute a calcium ion for a magnesium ion in the bulk of the crystal and at the surface, we shall call the calculated heat of segregation. This is shown in Figure 2 as a function of surface coverage. First, we note that there is a large negative segregation energy implying strong surface segregation. This is qualitatively what is observed and accords with our ideas on the lattice strain associated with a larger impurity ion. As previously mentioned, the calculated heat increases markedly with surface coverage, thus reducing the propensity for segregation. This is due to the increasing repulsion between neighbouring surface impurity ions and leads to a restructuring of the surface.

The experimentally measured heat of segregation is around −50kJ/mole at about 20% surface coverage of calcium. This does not seem to agree well with the calculation. However, the problem arises from the varying heat of segregation with surface coverage and the experimental definition of the heat from an Arrhenius plot. Figure 3 shows the experimental data and the calculated results plotted as the log of the surface ratio versus the inverse temperature. If the heat of segregation were constant, this would be a straight line whose slope is the heat (Section 3). In fact, the experimental results curve sharply as the temperature decreases. While this could be due to problems in equilibrating the system at low temperature, we note that the calculated results show

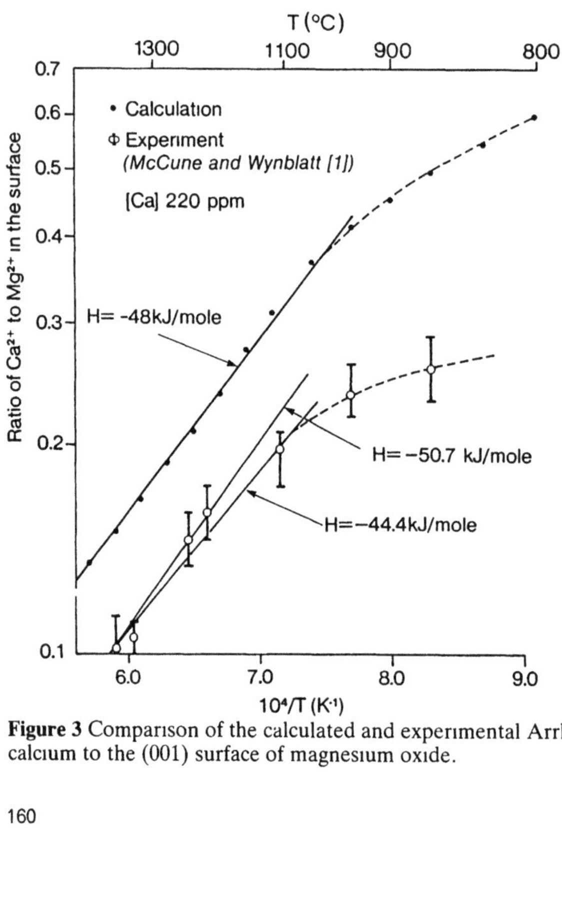

Figure 3 Comparison of the calculated and experimental Arrhenius plots for the segregation of calcium to the (001) surface of magnesium oxide.

a similar curvature due to the repulsion between the segregating ions. If we draw a straight line through the calculated data at high temperature, we deduce a heat of segregation of −48kJ/mole in good agreement with that measured. This provides the only sound way to compare experiment with calculation as a direct comparison of the calculated energies is inappropriate.

The only discrepancy between observation and calculation lie in the absolute value of the amount of segregation. This may be due to the neglect of entropy in this calculation but this is likely to be small, see below. The Auger results presented here average over an ill defined depth while the calculation considers the surface atomic layer explicitly. Recent experiments by McCune and Wynblatt /2/ using ion scattering have shown that the segregation is indeed very localised to the surface and consequently increased the measured surface concentrations. Thus, in this case, the experiment and calculation are in excellent quantitative agreement.

An interesting consequence of the high segregation energies observed in ionic materials is the corresponding large reductions in surface energy. The surface energy of an impure material can be considered as the energy to create new surface, for example by cleavage, while simultaneously allowing the new surface to equilibrate with the bulk impurity concentration through segregation. Thus the surface energy, for a low bulk concentration of a segregating impurity, is the surface energy of the pure material reduced by the segregation energy of the impurity multiplied by the equilibrium surface impurity concentration. When the segregation energy is large, this latter term can greatly reduce the surface energy at low temperature. Thus the low temperature surface energy of pure MgO is approximately $1Jm^{-2}$ but should be reduced to about $0.2Jm^{-2}$ if there is any significant calcium impurity content. For more strongly segregating impurities, the effect is even more dramatic. If MgO contains only 10ppm of barium impurity, we predict that the surface energy will be negative at all temperatures below the melting point /12/. This does not imply that such a crystal would fall apart. The segregation of an impurity cannot occur at a timescale comparable to the generation of new surface by fracture. However, it does suggest that such an impure crystal may be impossible to sinter since there is no thermodynamic driving force for the removal of surface.

3.3 Entropy

Although static lattice calculations only calculate the internal energy of the system, the free energy can be obtained by consideration of the lattice vibrations. If the phonons are calculated for the zero temperature lattice, the entropy and hence free energy can be determined within the harmonic approximation. However, one can do better than this by allowing for the lattice expansion with temperature. Calculation of the entropy via the phonons with a temperature dependent lattice parameter is known as the quasi-harmonic approximation. Of course, the lattice expansion should be calculated from the phonon pressure for consistency and the internal energy contribution evaluated at the appropriate lattice parameters also. When this is done, the calculated free energies, for defects in the bulk lattice, can agree closely with those obtained from molecular dynamic simulations and experiment /6/.

For surfaces, calculations have only been carried out so far within the harmonic approximation. This is because of the problems associated with maintaining constant volume when the surface must be free to relax. A method for calculating the surface entropy was suggested by DeWette, Chen and co-workers some years ago /15/. They considered a finite slab of crystal (e.g. 15 planes) that is periodic in two dimensions. The phonon dispersion curves for such a slab make a complicated pattern of lines derived mainly from the folding of the bulk phonon curves into the surface Brillouin zone but include the surface modes. The entropy can be calculated from the usual expression for an assembly of oscillators of frequency, ω_i.

$$S=k\sum_{i=1}^{N}\left\{-\ln(1-e^{-x})+\left[\frac{x}{(e^{-x}-1)}\right]\right\}$$

where $x=h\omega_i/kT$ and we assume a Boltzmann distribution for the occupancy of the phonon states. A similar, but simpler, calculation for a three dimensionally periodic lattice structure gives the bulk entropy with no surface contribution and so the difference gives the surface excess entropy.

Similar calculations can be carried out for the impure or defective crystal. Thus, the entropy of substitution of calcium in the bulk of MgO was calculated to be 0.3k while the equivalent surface value was determined to be only 0.06k /16/. The vibrational entropy of segregation is therefore calculated as −0.24k. This contributes to the free energy of segregation at 1000 K approximately 0.03eV per calcium ion and is an entirely negligible contribution, at least in this particular case. The major entropic contribution to segregation comes from the configurational terms which are, of course, included approximately in the expressions for the segregation isotherm.

3.4 Charged Defects and Impurities

When we consider the segregation of charged defects and impurities there are two additional complications. Firstly, since the species are charged, their interaction cannot be neglected except at the lowest concentrations. Thus the appropriate expression for the free energy must include the Coulomb interaction, and minimization of this gives the equilibrium surface concentration. This leads to the usual space-charge expressions for the intrinsic defects and the impurities interact with the surface charge also /17/. The solution of these equations for modest defect concentrations are well known and defect calculations can, in principle, calculate the appropriate energies. Secondly, when we calculate the energy of an isolated charged species near a crystal surface the terms in the energy are different from those for a neutral species. For a neutral defect, the energy can be described in terms of lattice strain. A charged defect, however, derives much of its stability from polarization of the surrounding lattice. Near a surface, there is less dielectric medium to polarize and so the energy increases. In classical electrostatics, this can be described by the method of images in which a charge approaches a dielectric interface. In the case of the free surface, the vacuum or gas has a lower dielectric than the host crystal. Thus, the image charge has the same charge as the defect and exerts a repulsive force. This is illustrated in Figure 4 which shows a calculation of a cation

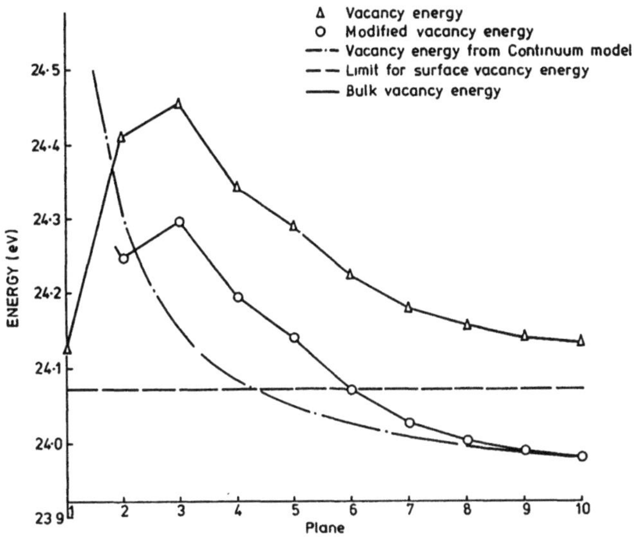

Figure 4 A calculation of the energy of a cation vacancy near the (001) surface of magnesium oxide. The calculation is compared with the electrostatic image charge limit (see text).

vacancy near the (001) surface of MgO. The x-axis shows the depth from the surface marked as atomic planes with plane 1 being the surface layer.

The figure shows two interesting features. The difference between the calculated vacancy energy and the modified vacancy energy comes from the surface dipole due to relaxation. The MgO surface rumples slightly at equilibrium with the oxygen ions moving outwards while the calcium ions move inwards /14/. This creates a small dipole that modifies the defect energies in the bulk of the crystal. After allowing for that, the defect energy is compared with the repulsive curve calculated from classical image theory. After a few planes, it agrees closely and tends only slowly (R^{-1}) to the bulk defect value (along the x-axis). Close to the surface, the lattice structure is important. In particular, the defect energy drops in the surface plane because there are fewer interatomic interactions to break when creating a surface vacancy rather than a bulk vacancy. This latter point leads one to expect surface vacancies to have a lower energy than the bulk energy but we see that the effect is cancelled by the opposing image term. In fact, the terms are well balanced so that in some cases surface vacancies have lower energy than the bulk while in other cases they are of higher energy.

Charged impurities also feel this image effect repelling them from the surface. One consequence is that the binding of impurities and defects into neutral clusters is greatly enhanced near the surface. A bulk defect binding energy of a fraction of an electron-volt can easily become 1–2 eV at a surface. The opposite effects should be observed at oxide-metal interfaces where the ionic material meets a material of much higher dielectric constant. Thus charged impurities should be stabilised near the interface and their mutual binding reduced.

3.5 Aliovalent Impurities

The most detailed study of the segregation of an aliovalent impurity in an ionic material is that of magnesium in α–alumina /3/, chosen because of its importance as a sintering agent. In the preliminary calculations of this system, advantage was taken of the strong surface binding energy between the impurity and vacancies /18/. Instead of explicitly treating the charged species and solving the necessary space-charge equations, we considered the segregation of neutral clusters of two impurity ions and an associated anion vacancy. As mentioned above, the segregation of even isovalent impurities in alumina is more complex than in magnesia because of the role of surface relaxation. Although we have simplified our problem to consider neutral clusters, we have the additional complication of the possible orientations for the cluster. Our study has concentrated on the basal plane (0001) of alumina since that was the subject of the experimental work. The lowest energy configuration at moderate surface coverages was found to consist of the magnesium substitutions in the surface layer and in the next repeat unit below with an oxygen vacancy in the uppermost oxygen plane. Full details of the configuration are given elsewhere /18/.

Figure 5 shows the calculated heat of segregation (i.e. the difference in the bulk and surface substitution energies) as a function of surface coverage of magnesium. Upto a quarter coverage, the calculated heat is approximately constant at −1.63eV and should give reasonable Arrhenius behaviour. After that there is a sharp decrease in the energy with a minimum at one third coverage, corresponding to the composition of spinel, $MgAl_2O_4$. The depth of the energy minimum is exactly the calculated heat of formation of spinel from magnesia and alumina. This behaviour translates into an Arrhenius plot which is linear initially and then abruptly changes to a constant higher coverage value corresponding to the formation of a surface phase of spinel. The calculated plot is shown in Figure 6 with the experimental points of Baik et al /3/.

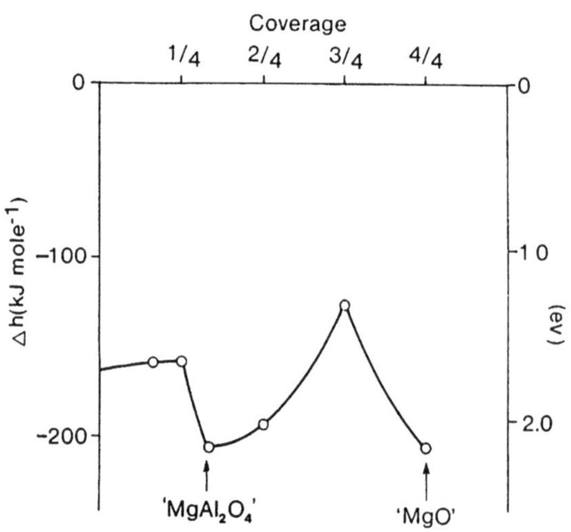

Figure 5 The calculated energy of segregation as a function of surface coverage for a magnesium impurity at the (0001) surface of α–alumina assuming neutral defect clusters

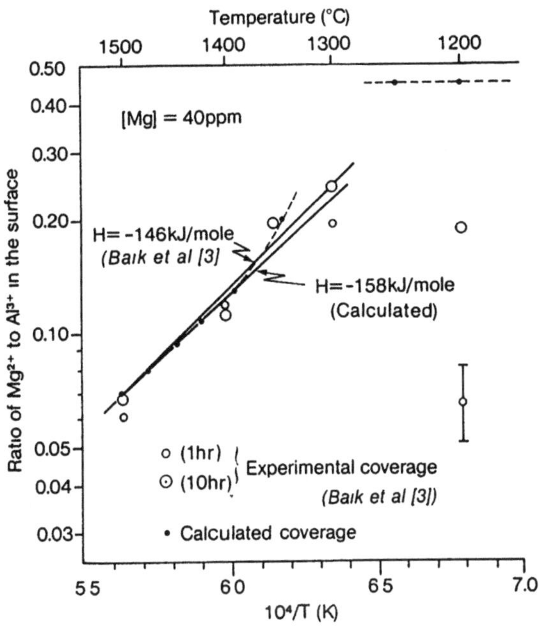

Figure 6 Experimental and calculated Arrhenius plots for the segregation of magnesium impurity to the basal plane surface of α–alumina

As before, uncertainties over the depth of the experimental resolution and in the defect vibrational entropy make comparison of the absolute concentrations difficult. Thus the experimental and theoretical results have been brought into alignment. We can see excellent agreement between the slopes of the Arrhenius plot at temperatures above about 1300 C. The calculated heat of segregation of -1.63eV agrees well with the measured value of -1.51eV. Note that in this case the calculated Arrhenius value is the same as the difference in the defect energies (Figure 5) because there is little variation of the heat with coverage in this region. At lower temperatures, we predict the abrupt spinel phase formation. Unfortunately, this occurs at the temperature where the experiments are unable to achieve equilibrium in a reasonable time, as shown in the plot. However, at lower temperatures, Baik et al did observe the appearance of extra reflections in the LEED pattern suggesting a change in surface structure.

4. Grain Boundary Diffusion

Our second example of the application of static lattice calculations to ionic materials is of diffusion along grain boundaries. The properties of grain boundaries are of particular importance in the study of oxide materials since they are almost invariably polycrystalline and often fine grained. Grain boundaries can determine much of the interesting behaviour such as sintering, electrical characteristics, mechanical properties and oxidation kinetics. The example chosen here is of grain boundary diffusion in nickel oxide.

Nickel oxide was chosen as a model experimental system for study by Atkinson and co-workers who were trying to understand the high temperature oxidation of metals in terms of the microscopic processes /19–21/. The growth of the oxide film on a metal often follows a $(time)^{\frac{1}{2}}$ dependence after the initial stages. This is interpreted in the Wagner theory as being due to rate limitation by either metal or oxidant transport through the growing scale. In the case of the oxidation of nickel, one expects the transport mechanism to be nickel diffusion through the oxide since NiO is non-stoichiometric with a nickel deficiency. Atkinson et al measured the diffusion rate through nickel oxide single crystals but found that it was too slow to account for the observed oxidation rate. They subsequently measured the grain boundary diffusion rate and found it substantially greater than lattice diffusion. When this was substituted into the Wagner theory it reproduced the observed scale growth kinetics. Recent experiments on specially prepared bi-crystals have failed to find enhanced diffusion along the grain boundaries but the reasons for this are not clear yet /22/. The calculations aim to see whether a simple theoretical model would predict enhanced grain boundary diffusion, to give a mechanistic interpretation and to study the influence of impurities on inhibiting the process.

4.1 Grain Boundary Structure and Properties

When this work was done, very little was known about the structure of any grain boundary in an ionic material. The first step was, therefore, to predict possible structures for nickel oxide. These were limited to range of pure tilt and pure twist boundaries of high periodicity. Such boundaries are known as coincidence interfaces and there is some debate in the literature as to whether they represent special low energy structures or are representative of more general boundaries. In any case, none of our structures can be considered to be typical of a boundary in the growing oxide film. Thus, we have studied several boundaries of different structure in the hope of deducing general behaviour that is not strongly dependent on the particular boundary configuration /23,24/. Some of the predicted structures show fairly open channels down the boundary plane which possibly suggest enhanced diffusion pathways while others are more close packed. Predicted structures for twist boundaries suggest a reduced density of ions in the boundary plane with substantial restructuring /25/. Most recently, high resolution electron microscopy has been used to obtain atomistic structures for the tilt boundaries in NiO /26/. The preliminary results reveal some similarities with the predicted structures but also some differences. More work is needed both experimentally and theoretically but it seems that we shall soon have detailed knowledge of grain boundary structures in this system. Similarly, X–ray diffraction studies may soon be able to give us the information needed to validate the twist boundary structures /27/.

These structures have been used to investigate the defect and impurity concentrations expected near grain boundaries /28/. Because of the lower symmetry structures, grain boundaries have a greater diversity of distinct sites than most free surfaces. Therefore, for each defect there is a range of energies in the boundary region. For every boundary studied, there is at least one site that is more favourable for any particular defect than the bulk. In nickel oxide, we considered cation and anion vacancies as well as isolated holes (Ni^{3+} ions) on four different boundaries. Vacancy segregation energies varied from −0.3 eV to −1.6 eV with boundary structure but were always negative. A more general (lower symmetry) boundary structure is likely to show the same behaviour since it will have an even greater diversity of sites. Thus we deduce that all defect concentrations will be enhanced near grain boundaries and this is likely to contribute to enhanced boundary diffusion. A space-charge calculation for the specific example of the $(2\bar{1}1)/[011]$ at 1000 K predicted a negatively charged boundary plane with a screening region with an enhanced hole concentration but a diminished cation vacancy concentration. Nevertheless, the vacancy concentration of the boundary region was enhanced overall by a factor of about 40 and the screening length was only about 2nm.

Similar calculations for impurities in NiO reveal the same behaviour. Because of the diversity of boundary sites, all impurities segregate to all grain boundaries. This is in contrast to the case for surfaces where some impurities segregate whilst others have a diminished concentration near the interface.

4.2 Migration

A static calculation attempts to determine the diffusion process by calculating the saddle point energy for the defect hop. This can then be used in a rate theory to deduce the absolute diffusion rate. This has been done with good results for some diffusion processes in the bulk lattice /29/. However, the complexity of these calculations has limited their application to the activation energies for diffusion. The total activation energy for diffusion is made up of the energy for defect hopping and the formation energy for the defect. Vacancies are created in an oxidation reaction,

$$\tfrac{1}{2}O_2 \rightleftharpoons O^{2-} + V_{Ni}'' + 2h \ ,$$

where the oxygen molecule dissociates and is absorbed in the lattice, leaving a vacancy and two holes on the nickel sublattice. Thus the activation energy, E_A, is given by,

$$E_A = E_H + E_O/3$$

where E_H is the hopping energy and E_O is the energy of the oxidation reaction given above. The results from this calculation are given in Table 1 /30/.

Although there is some variation in activation energy with boundary structure, all the boundaries studied should show enhanced diffusion. Even the coherent twin, $(1\bar{1}1)/[011]$, has a reduced activation energy for diffusion despite maintaining the full lattice density. The bulk activation energy, calculated at 2.9eV, is somewhat larger than the measured value of 2.5eV /19/. This may be due to the neglect of crystal field terms in the defect energy which should be included for transition metal oxides such as NiO. All the energies are similarly increased and the calculated energy ratios agree with the

Table 1. The calculated activation energy for cation vacancy diffusion down tilt boundaries in NiO. The last column is the ratio of the boundary to bulk activation energies.

Boundary	E_A (eV)	Ratio
(310)/[001]	2.08	0.71
(320)/[001]	1.62	0.56
$(2\bar{1}1)$/[011]	1.77	0.61
$(1\bar{1}1)$/[011]	2.23	0.77
Bulk	2.91	

measured ratio of 0.7 in polycrystalline NiO /20/. Similar relative reductions have been measured in the grain boundaries of a number of metals.

Similar analysis has been carried out for the diffusion of the minority oxygen species through nickel oxide /30/. The concentrations of oxygen vacancies and interstitials are determined by the Schottky and Frenkel equilibria while the majority cation vacancy concentration is fixed by the oxidation reaction. The activation energy for oxygen diffusion is then given by,

$$E_A^v = E_H^v + E_S - E_O/3$$

or

$$E_A^i = E_H^i + E_F - E_S + E_O/3$$

depending on whether the diffusing species is a vacancy or an interstitial. E_S is the Schottky energy and E_F is the anion Frenkel energy. The results for this calculation are given in Table 2. The boundary calculations indicate that diffusion of oxygen interstitials via the interstitialcy mechanism is the preferred mechanism as was previously found in the bulk /31/. The comparison with experiment, assuming the interstitialcy mechanism, is good. The calculations suggest that as for the cation vacancy the grain boundary diffusion should be enhanced over bulk diffusion.

Finally, experimental studies of ceria incorporated in the growing oxide film have shown that the growth rate is substantially reduced thus inhibiting corrosion /32/. Calculations indicate that substitutional cerium on the grain boundary will increase the activation energy for cation vacancy hopping by 0.7eV to approximately 2.5 eV. Since this is greater than the value for the bulk lattice it suggests that this could be the

Table 2. Activation energies for oxygen diffusion along one grain boundary and in the bulk of NiO. Columns 2 and 3 give the calculated values for vacancy and interstitialcy mechanisms and the last column shows the experimental result /33,34/.

	E_A^v	E_A^i	experiment
(320)/[001]	6.0	3.0	2.5
Bulk	9.1	5.5	5.3

mechanism for cerium inhibition of the scale growth. It derives from the strong interaction between the negatively charged vacancy and the positive substitutional impurity. The calculations also indicate that cerium is hardly soluble in nickel oxide, and questions remain as to how it is incorporated in the growing oxide.

Acknowledgement

Some of this work was carried out as part of the longer-term research within the Underlying Programme of the UKAEA.

2 P.Wynblatt and R.C. McCune, in Surfaces and Interfaces in Ceramic and Ceramic Metal systems, edited by J. Pask and A. Evans, p83, (Plenum Press, New York, 1981).
1 R.C. McCune and P. Wynblatt, J. Amer. Ceram. Soc. 66 (1983) 111.
3 S. Baik, D.E. Fowler, J.M. Blakely and R. Raj, J. Amer. Ceram. Soc. 68 (1985) 281.
4 C.R.A. Catlow and W.C. Mackrodt (eds), Computer Simulation of Solids, Lecture Notes in Physics, 166 (1982).
5 J.H. Harding, Phys. Rev., B32 (1985) 6861.
6 R.J. Tarento and J.H. Harding, J. Phys. C, 20 (1987) L677.
7 N.F. Mott and M.J. Littleton, Trans. Farad. Soc., 34 (1938) 485.
8 M.J. Sangster and A.M. Stoneham, Phil. Mag. B43 (1981) 597.
9 W.C. Mackrodt and R.F. Stewart, J. Phys. C, 12 (1979) 431.
10 E.A. Colbourn, W.C. Mackrodt and P.W. Tasker, J. Mater. Sci., 18 (1983) 1917.
11 W.C. Mackrodt and P.W. Tasker, to be published.
12 P.W. Tasker, E.A. Colbourn and W.C. Mackrodt, J. Amer. Ceram. Soc. 68 (1985) 74.
13 P.J. Lawrence, S.C. Parker and P.W. Tasker, (1987) in press.
14 P.W. Tasker, Adv in Cearmics, 10 (1984) 176.
15 T.S. Chen and F.W. DeWette, Surf. Sci. 74 (1978) 373.
16 P. Masri, P.W. Tasker, J.P. Hoare and P.W. Tasker, Surf. Sci. 173 (1986) 439.
17 K.L. Kliewer and J.S. Koehler, Phys. Rev. 140 (1965) 1226.
18 W.C. Mackrodt and P.W. Tasker, Mat. Res. Soc. Symp. Proc., 60 (1986) 291.
19 A. Atkinson and R.I. Taylor, Phil. Mag., A39 (1979) 581.
20 A. Atkinson and R.I. Taylor, Phil. Mag., A43 (1981) 979.
21 A. Atkinson, R.I. Taylor and A.E. Hughes, Phil. Mag., A45 (1982) 823.
22 F. Barbier, Thèse d'Etat, Université Paris–Sud, (1986).
23 D.M. Duffy and P.W. Tasker, Phil. Mag., A47 (1983) 817.
24 D.M. Duffy and P.W. Tasker, Phil. Mag., A48 (1983) 155.
25 P.W. Tasker and D.M. Duffy, Phil. Mag., A47 (1983) L45.
26 K.L. Merkle, J.F. Reddy, C.L. Wiley and D.J. Smith, in Interface Science and Engineering '87, edited by R. Raj and S.L. Sass, (1988) in press.
27 M.R. Fitzsimmons and S.L. Sass, in Interface Science and Engineering '87, edited by R. Raj and S.L. Sass, (1988) in press.
28 D.M. Duffy and P.W. Tasker, Phil. Mag., A50 (1984) 143 and 155.
29 J.H. Harding, Mat. Res. Soc. Symp. Proc., 60 (1986) 163.
30 D.M. Duffy and P.W. Tasker, Phil. Mag., A54 (1986) 759.
31 M.J.L. Sangster and D.K. Rowell, Phil. Mag., A44 (1981) 597.
32 A.T. Chadwick and R.I. Taylor, Solid St. Ionics, 12 (1984) 343.
33 A. Atkinson, F.C.W. Pummery and C. Monty, in Transport in Non-stoichiometric Compounds, edited by G. Simkovich and V.S. Stubican (Plenum Press, New York, 1985).
34 C. Dubois, C. Monty and J. Philibert, Phil. Mag., A46 (1982) 419.

Part IV

Nucleation at the Solid-Liquid Interface

Geometrical Structure of Adlayers on Immersed Electrodes

J.G. Gordon II[1], *O.R. Melroy*[1], *and L. Blum*[2]

[1]IBM Research Division, Almaden Research Center, 650 Harry Road, San Jose, CA 95120, USA

[2]Physics Department, University of Puerto Rico, Rio Piedras, PR 00931, Puerto Rico

Lead and copper monolayers (produced by underpotential deposition) on silver(111) and gold(111) surfaces, respectively, have been characterized by x-ray absorption spectroscopy and x-ray diffraction. The experiments were performed *in situ*, with the electrode immersed in aqueous electrolyte, and at a controlled potential. A combination of grazing incidence excitation and fluorescence detection (for x-ray absorption) provided the required sensitivity and surface selectivity. The metal adlayers are fully reduced (zero valent). Lead forms a well ordered two dimensional solid with a close packed hexagonal structure. Its lattice is slightly compressed relative to bulk lead and is incommensurate with the underlying silver lattice. The copper EXAFS is consistent with a commensurate overlayer with the copper atoms in the three fold sites. Oxygen atoms of the first layer of water or adsorbed electrolyte lie at a well defined distance from the metal adlayers.

1. INTRODUCTION

In discussions of dynamics one assumes, explicitly or implicitly, a specific underlying geometric structure. For the solid:vacuum interface, this structural picture is based on the results of fairly direct experiments, eg. RHEED, LEED, electron microscopy, or STM. For the metal:electrolyte interface, what is known is inferred from thermodynamic, kinetic, and optical measurements coupled with theoretical treatments based on highly idealized models which accurately describe only the diffuse double layer, or space charge region, within 10-100 Å of the electrode surface. While these measurements are sensitive to structure, and provide a reasonable picture of the interface, they cannot directly reveal structure at the atomic or molecular level.

In recent years there have been considerable advances in the development of new, primarily spectroscopic, probes of the solid:liquid interface.[1-9] These provide information about the identity and quantity of interfacial species and possibly their orientation, but do not directly reveal the geometrical arrangement of molecules at the interface. Another approach has been to apply ultrahigh vacuum (UHV) surface spectroscopic techniques.[10, 11] As these rely primarily on particles as probes, they are not effective in solution because the propagation distances are too short. For application, they require transfer of the electrode from the electrolytic medium into vacuum and therefore present the serious question of whether the examined surface is really identical to the surface in solution. Transfer from normal pressure (24 torr, the vapor pressure of water at 25°C) to 10^{-10} torr represents a destabilization (ΔG) by approximately 16 Kcal/mol for any surface structure stabilized by water. Since the solvent (water) is not a contaminant, but rather an essential component of the interface, one really needs to observe the structure *in situ* to have confidence in the results.

X-rays are the most promising probes to do this. X-rays with energies exceeding about 6 KeV do penetrate water and have wavelengths comparable to atomic spacings, so they can provide direct geometrical information. This article reports the use of *in situ* surface extended X-ray absorption fine structure (SEXAFS) and grazing incidence x-ray scattering (GIXS) to study metal adlayers under normal electrochemical conditions, i.e., under liquid electrolyte and at controlled potential. Two systems will be discussed, copper on Au(111) and lead on Ag(111).

Both of the above combinations exhibit underpotential deposition (UPD).[12] That is, the first layer of metal is deposited at potentials less negative than the formal thermodynamic deposition potential. In this initial deposition, the coverage changes from zero to saturation over a very narrow range and then remains constant until the potential for bulk deposition is reached. The experiments described below were performed at potentials in this region of constant coverage.

2. EXPERIMENTAL

As surface probes, X-rays have three major problems:

1. The scattering and absorption cross sections are small. For example, absorption by a monolayer of lead just above the L_{III} edge would result in a change in transmittance of $\Delta T/T = 5 \times 10^{-5}$.
2. They are not inherently surface selective, a consequence of their weak interaction with matter.
3. Scattering by "uninteresting" matter (which is present in much greater quantity than the interesting stuff) generates a large background signal.

We have used a combination of techniques to overcome these problems (See Fig. 1):

1. The exciting x-rays strike the sample at a grazing angle so that they are totally externally reflected. (At x-ray wavelengths, the refractive index of matter is less than one.) This restricts the penetration of the x-rays (the evanescent wave) to less than 50 Å.
2. The surface layer differs in elemental composition from the substrate (hence, its fluorescence line is at a different energy).
3. The x-ray flux is very high (synchrotron source).

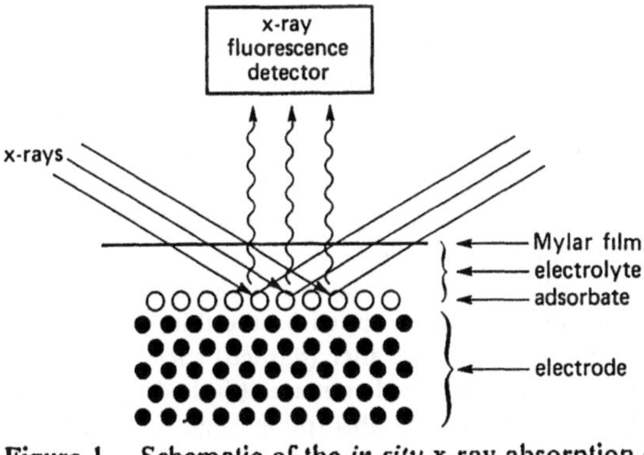

Figure 1. Schematic of the *in situ* x-ray absorption experiment. The angle of incidence is exaggerated. The scattering experiment is similar, except that the detector is moved.

4. The overlayers, electrolyte and window, are thin (< 50 μm), minimizing scatter.
5. For SEXAFS, the x-ray absorption is measured by detecting the characteristic fluorescence emitted by the excited atoms rather than by measuring a change in transmittance.
6. For GIXS, the incident beam is highly collimated, 1 mrad (synchrotron source).

Au(111) or Ag(111) electrodes were epitaxially grown by evaporation in vacuum onto a mica substrate held at elevated temperature.[13] This electrode formed one side of a thin layer cell. The other side was a 6.35 μm thick mylar sheet. When the electrode assembly was lowered into the electrolyte reservoir, the mylar film floated on the surface exposing the metal film to the bulk solution. The metal adlayers were deposited in this position. After deposition, electrolyte was withdrawn until just the bottom of the electrode was immersed. A thin layer of solution (estimated to be less than 50 μm thick) was held between the mylar film and the electrode surface by capillary action. In this way, the thin layer remained in electrical (and physical) contact with the electrolyte reservoir in which were situated the counter (Pt) and reference (Ag/AgCl, 3 M KCl) electrodes. Diffusion from the reservoir into the thin solution layer was very slow and even permitted polarization of the electrode negative of the formal deposition potential without significant additional metal deposition. The copper monolayer was deposited at +0.13 V from 5 x 10^{-5} M $CuSO_4$ in 0.5 M sulfuric acid solution. The lead monolayer was deposited at -0.53 V from 5 x 10^{-5} M lead acetate in a solution of 0.5 M sodium acetate and 0.1 M acetic acid.

For both the SEXAFS and GIXS experiments, the x-ray beam was incident through the mylar cover sheet at a grazing angle (< 0.5°). For EXAFS, the characteristic fluorescence lines were isolated by absorption filters (Ni for Cu and Ge for Pb) and an energy dispersive solid state detector (high purity Ge). The EXAFS data were analyzed in the standard way with interatomic distances being obtained by fitting the oscillatory part of the EXAFS equation to the experimental fine structure with phase shifts derived experimentally from model compounds.[14] Details of the cells and apparatus for both types of measurement are published elsewhere.[15, 16, 17]

3. RESULTS

3a. Cu on Au(111)

For these experiments, the gold electrode was poised at at +0.13 V in the region of saturated UPD coverage. From electrochemical and other measurements, the coverage at this potential is approximately one copper atom per surface gold atom and there is strongly suggestive evidence that the the copper layer is ordered and commensurate. To date, however, this has not been directly verified.[12, 18] The x-ray absorption spectrum of the copper adlayer was measured at the copper K-edge and is shown in Fig. 2. That this spectrum originates from reduced copper on the electrode surface was demonstrated by changing the potential to +0.5 V, where the UPD layer is oxidized to Cu^{2+}, and recording another spectrum. (The copper ions are trapped by the thin layer cell and, hence, remain in the x-ray beam.) Before (solid) and after (dashed) spectra are shown in the inset in Fig. 2. The spectrum of the oxidized material (dashed curve) differs significantly from that of the UPD layer. First, the absorption edge is shifted about 4 eV to higher energy, consistent with the increase in oxidation state, and suggesting that the UPD layer is fully reduced. Second, there is a strong resonance peak just above the

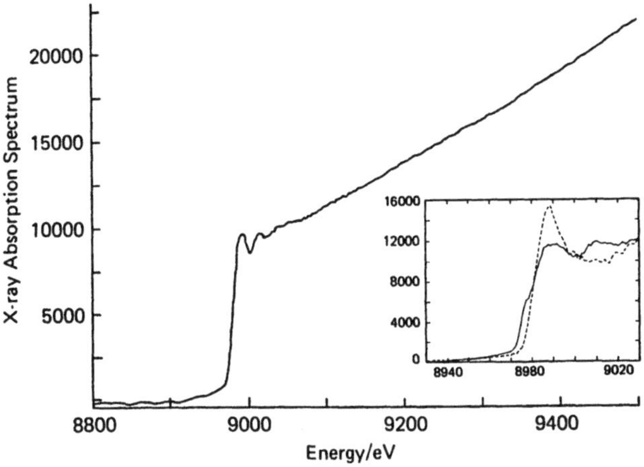

Figure 2. X-ray absorption spectrum at the Cu K_{α} edge for Cu on Au(111); electrode potential 0.13 V. Insert: Edge region at 0.13 V (dashed) and 0.50 V (solid).

edge which is absent in the spectrum of the UPD layer. This 'white line' is characteristic of Cu(II). Indeed the dashed spectrum is identical with that from an authentic copper sulfate solution.

The fine structure extracted from the absorption spectrum (Fig. 2) is shown in Fig. 3 along with the radial structure function (RSF) obtained by Fourier transformation. The dashed line in Fig. 3 is the filtered fine structure obtained by inverse transformation of the RSF between 1 and 4 Å.

Because the sample is a single crystal and the synchrotron x-ray beam is polarized, one can obtain information on orientation well as distance.[19] This is because the

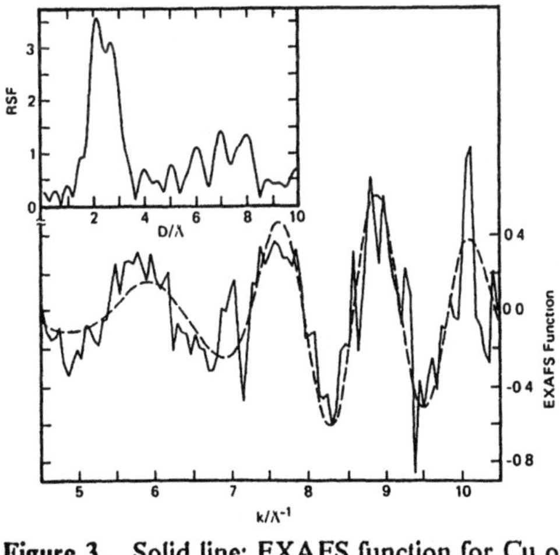

Figure 3. Solid line: EXAFS function for Cu on Au(111); Dashed line: Fourier filtered EXAFS (between 1 and 4 Å); Insert: Radial structure function (RSF) derived by Fourier transformation of EXAFS between k = 4 and 11 $Å^{-1}$.

contribution of a given absorber-scatterer atom pair to the total EXAFS intensity is proportional to the square of the dot product of the x-ray beam electric field (polarization) vector and the atom-atom vector, or

$$N_{eff} = 3 \sum_{j=1}^{N} \cos^2\theta_j,$$

where N_{eff} is the apparent number of atoms in the ith coordination shell, θ_j is the angle between the vectors and the summation is over all the atoms j in the ith coordination shell. Hence, atom pairs whose connecting vectors are parallel to the electric field vector (polarization direction) contribute significantly to the fine structure oscillations, whereas atom pairs whose connecting vectors are perpendicular to the polarization direction do not.

Accordingly, SEXAFS spectra were obtained in two different configurations: 1) with the x-ray beam polarized parallel [15] and 2) perpendicular [20] to the the plane of the electrode. The results are summarized in Table I. (The curves in Fig. 2 and 3 are for the parallel polarization.)

Table I. Interatomic distances for Cu on Au(111)

	↕	⟷
Cu-Cu	–	2.92±.03Å
Au-Cu	2.5±.06Å	2.58±.03Å
Cu-O	2.05±.03Å	–

The spectra for the parallel polarization yield two peaks in the RSF corresponding to copper-gold and copper-copper scattering. The copper-copper distance is identical within experimental error to the gold-gold distance (2.89 Å) in the (111) surface. Furthermore, copper-copper scattering is observed only for parallel polarization indicating that all the copper atoms lie in the same plane. Copper-gold scattering is observed in both polarizations. This means that the copper atoms do not occupy atop sites. There is no evidence for more than one copper-gold distance, indicating that the copper atoms occupy a single symmetrical site. The amplitude of the copper-gold scattering is consistent with the three fold hollow as the adsorption site. Considered in combination with the coverage, these data indicate that the copper adlayer is commensurate with the gold lattice and that the copper atoms occupy three fold hollows.

Surprisingly, the RSF from the SEXAFS spectrum in perpendicular polarization has a second peak at a shorter distance than copper-gold. Judging from the shape of the amplitude function, it involves scattering from a light atom, presumably oxygen. If this assignment is correct, and since copper oxide is not stable in acid, this oxygen must be from interfacial water or adsorbed sulfate. Furthermore, this scattering is observed only in perpendicular polarization, so this atom sits directly above a copper atom. We believe that the most likely candidate is adsorbed sulfate, and therefore propose that the local site geometry about a copper atom in the UPD layer on gold(111) is as depicted in Fig. 4.

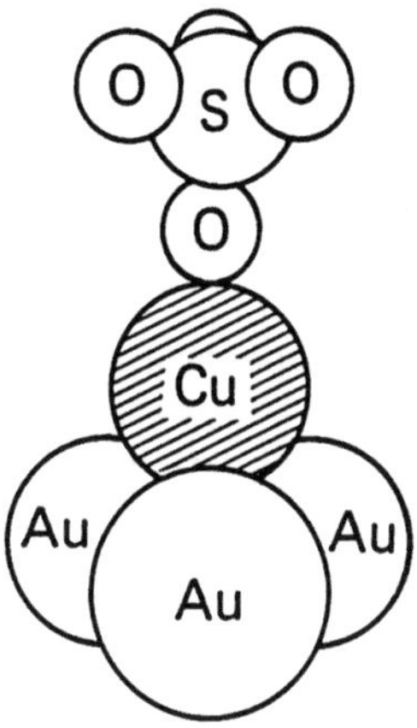

Figure 4. Local site geometry around copper.

3b. Pb on Ag(111)

The lead layer was studied by both SEXAFS [16] and surface diffraction [17, 21] at -0.53 V which is in the region between underpotential and bulk deposition. The coverage here, evaluated from the total charge transferred, is 1.6×10^{-9} mol/cm^2 (9.4×10^{14} atom/cm^2), close to that expected for a hexagonal close packed layer with bulk lead spacing. Reflectance [7] and ellipsometry [8] measurements support this interpretation, but there has been no direct structural determination. The X-ray absorption spectra were recorded at the Pb L_{III} edge with the x-ray beam polarized perpendicular to the electrode surface. As above, comparison of the spectra before and after oxidation of the adlayer demonstrated that the analyzed spectrum arose from a reduced lead layer on the silver surface. Analysis of the EXAFS reveals only one distance and it is too short to be Pb-Ag. Furthermore, the amplitude function is compatible only with scattering from a first row element. Given the composition of the solution, the most likely candidate is oxygen, from either water or acetate. With this assignment, a Pb-O distance of 2.33±0.02 Å is obtained. This is longer than in PbO_2 (2.16-2.22 Å) or red PbO (2.30 Å) and is consistent with coordination to lead in a more reduced oxidation state.

An EXAFS spectrum was also obtained at an electrode potential of -1.0 V. Even though this is far negative of the reversible lead potential, the thinness of the cell limited diffusion so that there was no appreciable further deposition of lead during the data collection. The spectrum closely resembles that at -0.53 V. Analysis yields a single Pb-O distance of 2.38±.02 Å. No Pb-Ag scattering was detected. The Pb-O distance is 0.05 Å longer than at -0.53 V, and is consistent with repulsion of a charged or, at least, strongly dipolar adsorbate by a more negatively charged electrode.

The apparent absence of Pb-Ag scattering requires comment. It is probably the result of a large Debye-Waller factor for lead and the incommensurate nature of the adlayer. Large thermal motion is expected from lead's low melting point (600 K). Also, the melting point of Pb monolayers on Cu single crystals is reported to be even lower than the bulk value.[22] It is reasonable to expect Pb to exhibit similar behavior on Ag. Support for this view comes from the observation that the x-ray absorption spectrum of lead foil obtained during this study did not exhibit any fine structure. Furthermore, the

Pb adlayer is certainly not commensurate with the underlying Ag lattice because of the large difference between the atomic radii of Pb and Ag (1.75 and 1.44 Å, respectively). The saturation coverage, given above, is consistent with a hexagonal close-packed monolayer with the normal Pb-Pb spacing. The effects of thermal motion and an incommensurate structure are similar and are not possible to distinguish from EXAFS data alone.

In this EXAFS experiment, the X-ray beam was polarized perpendicular to the electrode surface so information could not be obtained on the Pb-Pb spacing which would lie in the plane of the electrode and hence perpendicular to the E vector. This in-plane structure was probed by x-ray diffraction.[17]

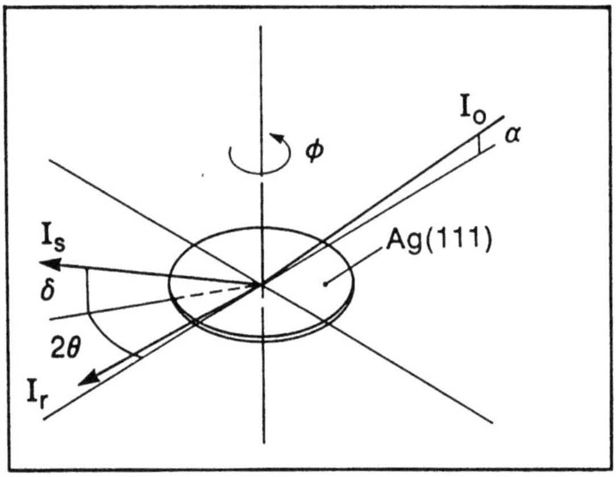

Figure 5. Scattering geometry: α is the incident angle, δ the output angle, θ the scattering angle, and ϕ the azimuthal angle of the sample.

Figure 5 illustrates the scattering geometry and defines the necessary angles. Figure 6 shows a radial (θ - 2θ) scan for the (10) reflection. This peak was confirmed to result from the lead monolayer by observing its disappearance when the electrode was poised at 0 V, at which potential, lead is oxidized to Pb^{2+} and dissolves in the solution. The relatively large background is due primarily to diffuse scattering from the thin layer of solution covering the electrode. Identical diffraction peaks are observed at 60° intervals, consistent with the six fold symmetry of a hexagonal layer. The (11) and (20) reflections were also observed, each with the expected six fold symmetry. The in-plane nearest neighbor distance, calculated from the diffraction angle, is 3.459 Å. This represents a 1.2% contraction from the distance in bulk lead (3.501 Å). From the full width half maximum (FWHM) of the peak, 0.037 $Å^{-1}$, the lead domains are estimated to be >150 Å in diameter. This is comparable to that observed for a lead monolayer on copper (110) in UHV.[23]

Actually, there are two equivalent orientations of the lead adlayer on the silver surface as can be seen from Fig. 7, which shows an extended rocking (ϕ) scan of the (10) reflection. A straight line background has been subtracted from this data. The are two peaks rotated $\pm 4.4°$, respectively, from the Ag $(\bar{2}11)$ direction, which is defined as $\phi = 0$. A similar pair of peaks rotated $\pm 4.4°$ from the corresponding silver direction

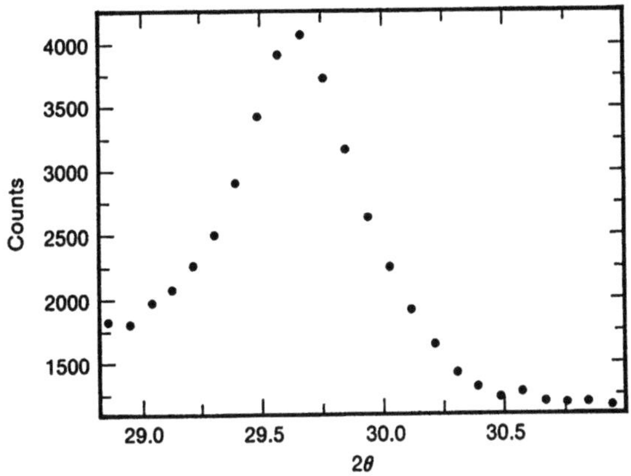

Figure 6. Radial scan at $\phi = 4.4°$ and $\alpha = \delta = 0.54°$

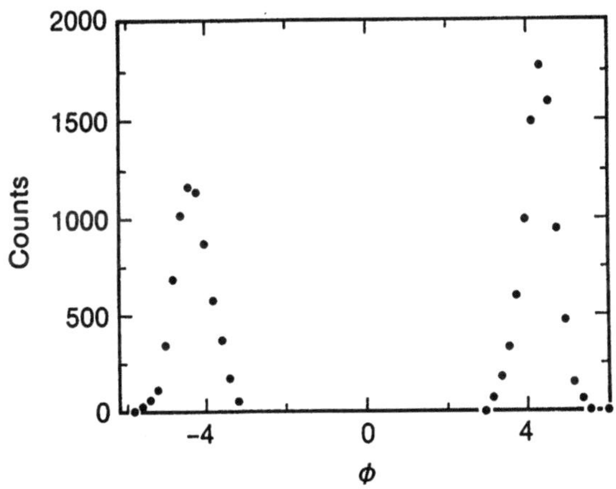

Figure 7. Rocking scan of (10) reflection of Pb on Ag(111) at $2\theta = 29.688°$ and $\alpha = \delta = 0.54°$.

is observed for each of the other reflections. The difference between the intensities of the +4.4° and -4.4° are real. We do not know its origin, but presume that it is caused by an anisotropic nucleation process. A similar difference was observed for O_2 and D_2 on graphite using LEED.[24]

Figure 8 illustrates the proposed structure of UPD lead on a silver (111) surface. The lead overlayer is ordered and epitaxial, but incommensurate with the silver lattice, and is rotated with respect to the silver lattice by 4.4°. This rotation angle is less than the 6.5° predicted by the McTague and Novaco model[25] but is in close agreement with that observed by Takayangi *et al.*[7] in their LEED study of vacuum deposited lead on silver (111).

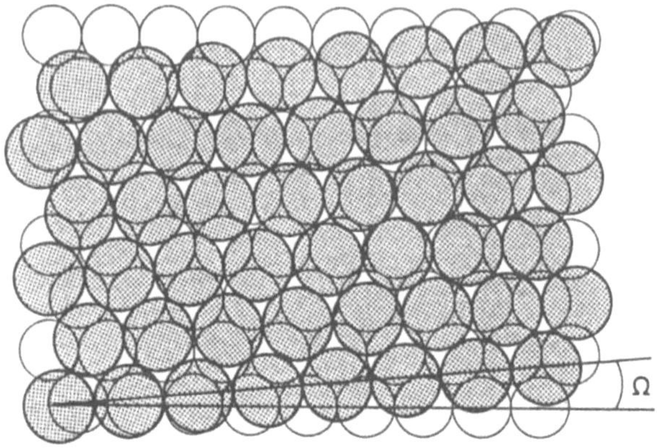

Figure 8. Proposed structure of Pb monolayer. Open circles represent the substrate silver atoms and shaded circles the overlying lead atoms.

4. DISCUSSION

SEXAFS and GIXS are complementary techniques, one probing local geometry and the other long range order. Together, they create a fairly detailed and unambiguous picture of the structure of UPD Pb on Ag(111). At full coverage, the Pb layer is hexagonally close packed, incommensurate with the substrate, and compressed 1.2% with respect to bulk metal. Acetate anions (probably) are adsorbed to the lead layer through oxygen atoms. X-ray scattering is not so easily applied to Cu on Au(111) since the adlayer diffraction peaks would coincide with those of the substrate and only small intensity differences would differentiate the bare from the adsorbate covered surface. Nevertheless, making use of the polarization dependence of the SEXAFS spectrum provides a quite detailed structural picture. An oxygen atom, probably from adsorbed sulfate, sits directly atop the Cu atom. The Cu atoms occupy three-fold hollow sites and form a a (1 x 1) commensurate superlattice with the gold substrate. Although the SEXAFS cannot provide direct proof of long range order, it is hard to envisage another model which will fit all of the data, especially in light of the supporting evidence from the *ex situ* RHEED data of Nakai *et al.*[18]

These two examples demonstrate that x-ray techniques can yield unambiguous structural information from immersed surfaces and open up the realm of *in situ* investigation of interfaces of electrochemical interest. They also show that well-ordered surfaces do exist under solution, despite the high concentration of water molecules. Furthermore, at least in these two specific cases, the adlayers have structures very similar to those observed for the same pair of metals in vacuum.

5. ACKNOWLEDGEMENTS

This work was partially supported by the Office of Naval Research and experiments were performed at the Stanford Synchrotron Radiation Laboratory (SSRL) which is supported by the Department of Energy and the Cornell High Energy Synchrotron Source (CHESS) which is supported by the NSF.

References

1. Proc. Int. Conf. on Non-Traditional Approaches to the Study of the Solid Electrolyte Interface, Eds., T.E. Furtak, K.L. Kliewar, and D.W. Lynch, Surf. Sci., 101 (1980)
2. R. Chang and T. Furtak: Surface Enhanced Raman Scattering, (Plenum Press, New York, 1982)
3. M. Weaver, F. Barz, J. Gordon and M. Philpott: Surface Science, 125, 409 (1983)
4. K. Kinimatsu, H. Seki, W. Golden, J. Gordon and M. Philpott: Langmuir, 2, 464 (1986)
5. R. Corn, M. Romagnoli, M. Levenson and M. Philpott: J. Chem. Phys., 81, 4127 (1984)
6. G. Richmond: Langmuir, 2, 132 (1986)
7. K. Takayanagi, D. Kolb, K. Kambe and G. Lehmpfuhl: Surface Science, 100, 407 (1980)
8. R. Muller and J. Farmer: Surface Science, 135, 521 (1983)
9. J. Gordon and S. Ernst: Surface Science, 101, 499 (1980)
10. D. Kolb: In Advances in Electrochemistry, H. Gerischer and C. Tobias, Eds., Vol II, (Wiley, New York, 1978) p.125
11. A. Hubbard: Acc. Chem. Res., 13, 177 (1980)
12. D. Dickertmann, F. D. Koppitz and J. W. Schultze: Electrochim. Acta 21, 967 (1976)
13. K. Reichelt and H. O. Lutz: J. Cryst. Grow., 10, 103 (1971)
14. P. A. Lee, P. A. Citrin, P. Eisenberger, and B. M. Kincaid: Rev. Mod. Phys., 53, 769 (1981)
15. O. R. Melroy, M. G. Samant, G. L. Borges, J. G. Gordon II, L. Blum, J. H. White, M. J. Albarelli, M. McMillan and H. D. Abruna: Langmuir, in press
16. M. Samant, G. Borges, J. Gordon, L. Blum and O. Melroy: J. Am. Chem. Soc., 109, 5970 (1987)
17. M. Samant, M. Toney, G. Borges, L. Blum, and O. Melroy: Surf. Sci., 193, L29 (1988)
18. Y. Nakai, M. S. Zei, D. M. Kolb, and G. Lehmpfuhl: Ber. Bunsenges. Phys. Chem., 88, 340 (1984)
19. J. Stohr, R. Jaeger, S. Brennan: Surf. Sci., 117, 503 (1982)
20. L. Blum, H. Abruna, J. White, J. Gordon, G. Borges, M. Samant and O. Melroy: J. Chem. Phys., 85, 6732 (1986)
21. M. Samant, M. Toney, G. Borges, L. Blum, and O. Melroy: J. Phys. Chem., in press
22. J. Henrion and G. E. Rhead: Surf. Sci, 29, 20 (1972)
23. S. Brennan, P. H. Fuoss, and P. Eisenberger: Phys. Rev. B, 33, 3678 (1986)
24. M. Toney and S. Fain: Phys. Rev. B, 36, 1248 (1987); J. Chui and S.C. Fain: J. Vac. Sci. Techn., A5, 710 (1987)
25. J. P. McTague and A. D. Novaco: Phys. Rev. B, 19 (10), 5299 (1979)

Electrochemical Processes in Small Systems

M. Fleischmann

Department of Chemistry, University of Southampton,
Southampton, Hants SO9 5NH, England

1. General Introduction to Electrode Processes

Electrochemical reactions involve the transfer of charges (ions or electrons) across the interfaces between two phases and the systems studied now include examples of nearly all the possible combinations of phases (e.g. solid-solid, solid-liquid, solid-gas, liquid-liquid, liquid crystalline-solid, etc.) and materials (metals, semiconductors, ionic conductors, polymers, insulators). However, the most widely studied group remains that of the interfaces between metals and ionically conducting solutions and most of the examples to be discussed here belong to this group. A widely accepted model of the structure of the interfacial region of such systems is illustrated in fig. 1; [1]; such models have been derived in the main from thermodynamic studies of perfectly polarised interfaces (interfaces which do not support sustained Faradaic processes). A potential difference $(\phi_m - \phi_s)$ is established over a distance of molecular dimensions and, if the magnitude of this potential difference exceeds that corresponding to the value $(\phi_m - \phi_s)_r$ corresponding to thermodynamic equilibrium, a sustained Faradaic process becomes possible. The driving force of electrochemical reactions

$$\pm \eta F = [(\phi_m - \phi_s) - (\phi_m - \phi_s)_r]F \qquad (1)$$

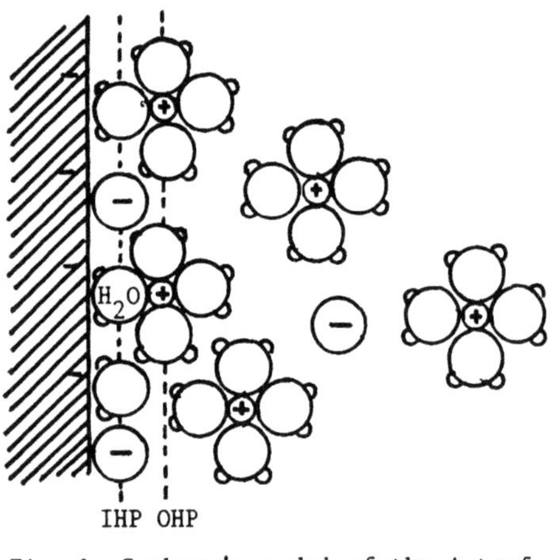

Fig. 1 Grahame's model of the interfacial region in the immediate vicinity of a metal electrode. IHP, the inner Helmholtz plane, is the distance of closest approach of unsolvated anions; OHP, the outer Helmholtz plane, is that of solvated cations. The solvent (here water) may be free or involved in primary solvation.

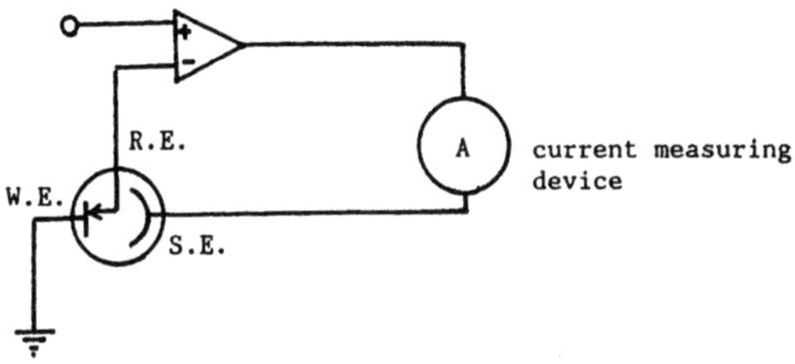

Fig. 2 A simple potentiostat. W.E. is the working, S.E. the subsidiary and R.E. the reference electrode.

can be as large as 3-4 eV; this is comparable to the strengths of chemical bonds and this accounts for the wide range of reaction types which have been observed. η is called the overpotential. The change in the potential difference at the interface is monitored by using an unpolarised reference electrode in conjunction with a capillary probe, fig. 2. The electrochemical cell can be made part of a feed back circuit (a potentiostat) and the response at the interface to a variety of stimuli (potential steps, linear sweeps in potential, small amplitude A.C. signals) is monitored by the current response. It is of interest that the first example of the application of relaxation techniques, a study of the behaviour of adsorbed hydrogen on platinum, [2], was interpreted in terms of the surface diffusion of hydrogen. An example of linear sweep voltammetry, the behaviour of platinised platinum in acid solutions, is illustrated in fig.3. We distinguish three regions: surface oxide formation in region C, the reduction extending into the perfectly polarised region B in such dynamic experiments and, finally the formation of at least two types of adsorbed hydrogen at potentials positive to the reversible hydrogen potential in region A i.e. at an underpotential. The analogy between linear sweep voltammetry (whose beginnings also date back some 60 years! [3-6]) to thermal desorption measurements will be apparent. The

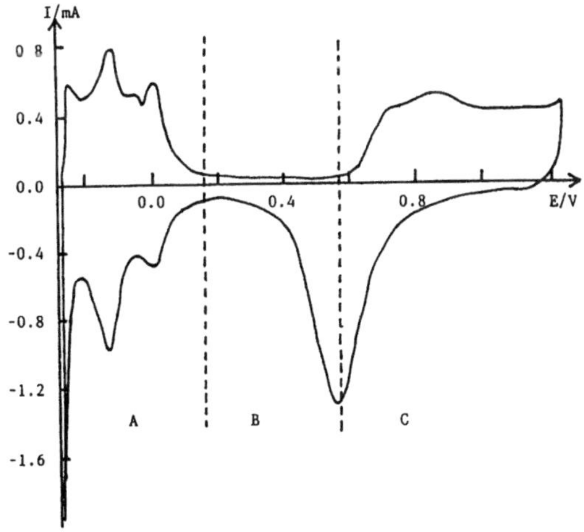

Fig. 3 Typical voltammogram of a platinized Pt electrode used in these investigations. Solution: 1M H_2SO_4. Apparent surface area ~ 0.41 cm^2. Potentials versus a saturated calomel reference electrode (S.C.E.)

major differences between the two types of studies are firstly, the possibility of the application of repeated fast sweeps in the case of electrochemical systems and, secondly, the use of the potential difference at the surface as an extra experimental variable. Furthermore, as electrochemical processes automatically transduce the rates of reaction into electrical signals, it becomes straightforward to make measurements on very small systems (current research deals with electrode areas down to 10^{-10}-10^{-9} cm^2 [7]; areas as small as 10^{-14} cm^2 can be investigated in special situations, see below) and/or on short time scales (typically 1 μs -1 ms; 10 ns is achievable by using rectification techniques [8,9]). The possibility of repetitively modulating the system between two (or more) states allows a virtually unlimited extension of the sensitivity. This is of particular importance to the application of <u>in-situ</u> spectroscopic and diffraction techniques to derive molecular level information about the interfaces as much methods can be made surface selective, where necessary, by modulating the system between two states where defined changes take place, e.g. the electrode with and without an adsorbate such as regions A and B of fig. 3 (see also final section).

Although a large proportion of electrochemical studies deals with the interconversion of soluble reactants and products and the formation or reaction of gases, attention here will be confined to the rôle of diffusion, to adsorption and to nucleation and phase growth processes since these topics will be of primary interest to participants in this series of meetings.

2. The Role of Diffusion in Electrochemical Reactions

The participation of diffusion of species through the solution (and/or the electrode) must always be taken into account. The kinetics of an interfacial reaction are governed by a rate constant which in the simplest cases will have the form

$$k = k^{\ominus} \exp\left(-\frac{\alpha\eta^{\ominus}F}{RT}\right) \tag{2}$$

where $k^{\ominus}$ is the rate constant at the standard potential and $\eta^{\ominus}$ is the overpotential measured on that scale. Thus kinetically controlled processes will be observable provided

$$k < k_m \simeq \frac{D}{\delta_c} \quad (\mathrm{cm\ s^{-1}}) \tag{3}$$

where k_m (cm s^{-1}) is the mass transfer coefficient, D (cm^2 s^{-1}) is the diffusion coefficient through the solution and δ_c (cm) is the thickness of the concentration boundary layer at a planar electrode. Even in well stirred solutions δ_c is minimally 10^{-3} cm so that the maximum value of k which can be resolved in quasi-stationary state experiments is $\sim 10^{-2}$ cm s^{-1}. By contrast, in the application of relaxation techniques

$$\delta_c \sim (D\tau)^{\frac{1}{2}} \quad (\mathrm{cm}) \tag{4}$$

where τ(s) is the characteristic time scale of the experiments. Thus for say τ = 10 μs, k_m can reach values as high as 1 cm s^{-1}. In recent years the attainment of high rates of mass transfer have increasingly been sought by using very small systems. Such micro electrodes (e.g. discs) rapidly become surrounded by quasi-spherical diffusion fields and k_m reaches the value

$$k_m \sim \frac{D}{a} \quad (\mathrm{cm\ s^{-1}}) \tag{5}$$

where a (cm) is the characteristic dimension of the microelectrode. An important feature of experiments using such systems is that the high rates of mass transfer (typically up to 1 cm s^{-1} for disc electrodes; values as high as 50 cm s^{-1} can be reached with thin ring electrodes) are achieved in the steady-state (diffusional relaxation time $\simeq 4a^2/D$(s)) so that the inherent complications of fast transient techniques are avoided [10].

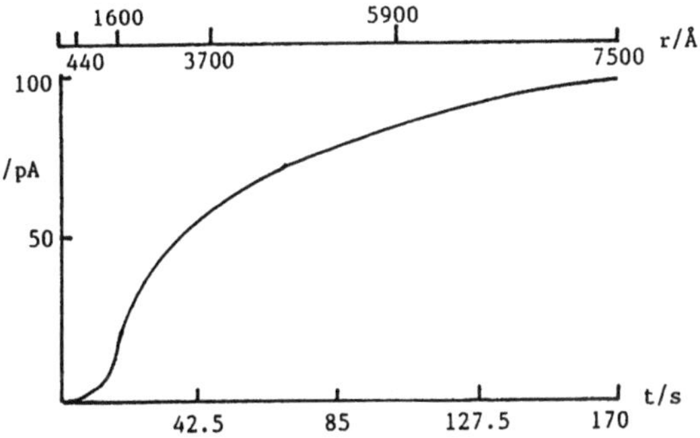

Fig. 4 Current-time transient for the deposition of a single droplet of mercury from a solution 0.2 mM $Hg_2(NO_3)_2$ + 10 mM NHO_3; overpotential 5 mV; 4μ radius carbon disc substrate electrode; k^{σ} derived from (6) using $\alpha = 0.5$ is 0.0057 cm s^{-1}. Radius of Hg droplet determined by integration of the i vs t transient.

These aspects are well illustrated by the electrodeposition of a single mercury droplet onto a carbon microelectrode [11], fig. 4. The small area of the substrate serves to restrict nucleation to a single growth centre and, although the deposition of mercurous ions onto mercury is one of the fastest electrochemical reactions known, the high rate of diffusion to the small growth centre ensures that the reaction is initially kinetically controlled (current proportional to the surface area of the growth centre) so that (compare [12,13])

$$i = \frac{32\pi M^2 F}{\rho^2} \left\{ k^{\sigma} \, (C^{\infty})^{(1-\alpha)} \, (C^{\sigma}) \exp^{\alpha}\left(-\frac{2\alpha\eta F}{RT}\right) \times \left[1 - \exp\left(\frac{2\eta F}{RT}\right)\right] \right\}^3 t^2 \tag{6}$$

Here M (g mol^{-1}) is the atomic weight of mercury, ρ (g cm^{-3}) is the density and C^{∞} and C^{σ} (moles cm^{-3} are the bulk and standard concentrations of the mercurous ions in solution respectively. At longer times, as the centre grows, the process becomes diffusion controlled and

$$i = \frac{2\pi\rho F}{M} \left\{ \frac{4MDC^{\infty}}{\rho} \left[1 - \exp\left(\frac{2\eta F}{RT}\right)\right] \right\}^{3/2} t^{\frac{1}{2}} \tag{7}$$

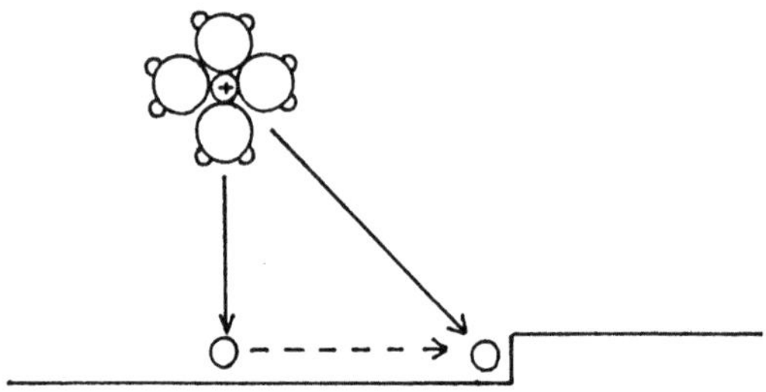

Fig. 5 Bulk ——— and surface ----- diffusion of ions and adatoms to an edge site.

Phase growth due to such fast electrode reactions has usually been studied under these diffusion controlled conditions e.g. [14,15] (see further below).

The participation of surface diffusion of adatoms has also frequently been invoked in the modelling of electrochemical reactions such as in phase growth by the propagation of steps of monatomic height generated by the classical mechanism of two-dimensional nucleation and growth or by the rotation of emergent screw dislocations, fig. 5. The participation of surface diffusion can certainly be detected in particular reactions (e.g. see the electrocrystallisation of nickel [16] below) or by special experiments designed to monitor the process. However, there is as yet little experimental evidence for the contribution of such surface diffusion to most overall reaction mechanisms and, indeed, this particular reaction path is unlikely for two reasons. In the first place, electrode reactions always require the desolvation of ionic species and will therefore be appreciably retarded as compared to comparable processes at solid-gas interfaces i.e. the slow steps in the overall reaction sequences will usually be chemical steps at the surfaces, e.g. of lattice formation at lattice forming sites in the case of phase growth. Secondly, if surface diffusion is retarded, then the parallel reaction path of rapid diffusion through the solution to the lattice growth sites will become predominant. Thus surface diffusion will either be difficult to observe or else be bypassed by diffusion through the bulk of the solution. It should also be noted that, taking again the example of electrochemical phase growth, the lattice forming sites propagate into regions of high concentration. For the particular example of the propagation of the edges of layer planes a very high rate of steady state diffusion through the solution is established [17] (k_m several hundred cm s^{-1}) so that the reactions remain controlled by lattice formation at the interface even for highly forcing conditions.

3. Nucleation

The nucleation of two- and three-dimensional growth centres by electrodeposition reactions [18] is entirely analogous to other examples of nucleation. Thus the electrochemical free energy of a spherical cluster of radius r is

$$\Delta\bar{g} = \pm \frac{4\pi r^3 \rho F \eta n}{3M} + 4\pi r^2 \sigma \qquad (8)$$

where the positive sign applies to cathodic processes (η negative) and the negative sign to anodic processes (η positive), n is the number of Faradays transferred per mole of reaction and σ (ergs cm^{-2}) is the surface energy. Thus on this 'classical' view of nucleation $\Delta\bar{g}$ passes through a maximum

$$\Delta\bar{g}_{max} = \frac{16\pi M^2\sigma^3}{3\rho^2 n^2 F^2 \eta^2} \tag{9}$$

which acts as the free energy of activation of the process. The frequency factor for the propagation of clusters through the population can be calculated with the usual drastic approximations (mainly by restricting attention to the region of the critical nucleus). Although the functional form of (9) has been observed for electrochemical nucleation ($1/\eta^2$ dependence), the sizes of the critical nuclei derived are invariably small. Nucleation processes are therefore frequently interpreted in terms of the 'atomistic' theory [19] which restricts attention to the aggregation of a limited number of species to form the critical nucleus.

For nucleation on a microelectrode, the onset of the growth of a single growth centre marks the time of birth [14,15] and the distribution of the arrival times is determined by the birth and death rates governing the formation of a single cluster, fig. 6:

$$\frac{dP_j}{dt} = \lambda_{j-1}P_{j-1} + \mu_{j+1}P_{j+1} - (\lambda_j + \mu_j)P_j \tag{10}$$

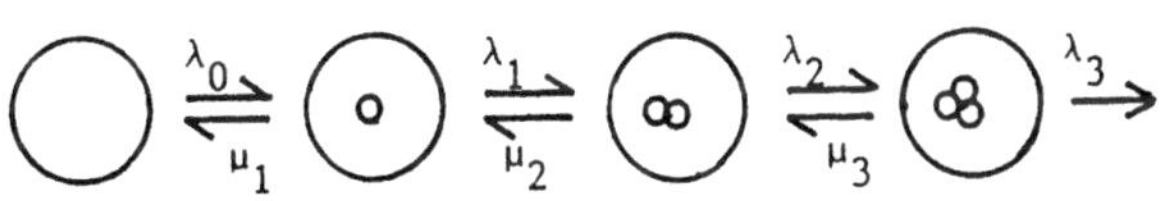

Fig. 6 The birth and death processes for the formation of a single cluster on a microelectrode.

Here we assume that the state # k+1 will persist once it is formed; when $j > k$, the cluster grows spontaneously and $\lambda_k \gg \lambda_{k-1}$, $\mu_{k+1} \simeq 0$ so that #k is the critical nucleus which is an absorbing state. The difficulties of obtaining general solutions to such sets of equations are well known: even for simple models of the clustering processes λ_j, μ_j are complicated functions of j. The problem can, however, be considerably simplified if we investigate the nucleation at very high overpotentials of a rigid lattice (electrochemical processes can be followed under highly forcing conditions). For such systems and conditions it is then likely that nucleation is a pure birth process and that the λ_j are closely similar. Here we will explore the limiting case where the λ_j for all clusters smaller than the critical nucleus j = k are identical, fig. 7:

$$\frac{dP_j}{dt} = \lambda\,(P_{j-1} - P_j) \tag{11}$$

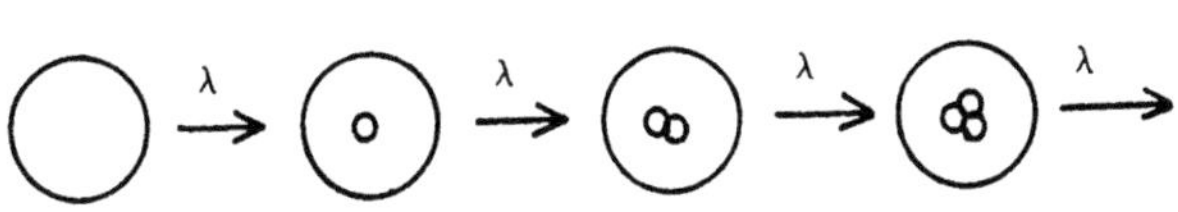

Fig. 7 Formation of a single cluster on a microelectrode by a pure birth process.

The set (11) has the well known general form of the Poisson distribution law

$$P_k = \frac{(\lambda t)^k}{k!} \exp(-\lambda t) \tag{12}$$

Thus the probability of forming a cluster of size greater than k in a given interval (0, t) is given by

$$\sum_{k+1}^{\infty} P_j = 1 - \sum_{0}^{k} \frac{(\lambda t)^j}{j!} \exp(-\lambda t) \tag{13}$$

Fig. 8 illustrates the form of the probability distributions for a number of values of k. Changes in the assumptions (variation of λ with j, inclusion of the μ_j) in general lead either to a broadening of the distributions or to changes in the interpretation of the dimensionless variable, T [20].

Fig. 9 illustrates the birth of a single growth centre of α-PbO_2 on a carbon microelectrode at a high overpotential. In view of the hemispherical form of the growth centres the kinetics follow the rate law [20,21].

$$i = \frac{2\pi nFM^2k^3(t - t_o)^2}{\rho^2} \tag{14}$$

where k (mol cm^{-2} s^{-1}) is the rate constant of crystal growth, t_o is the

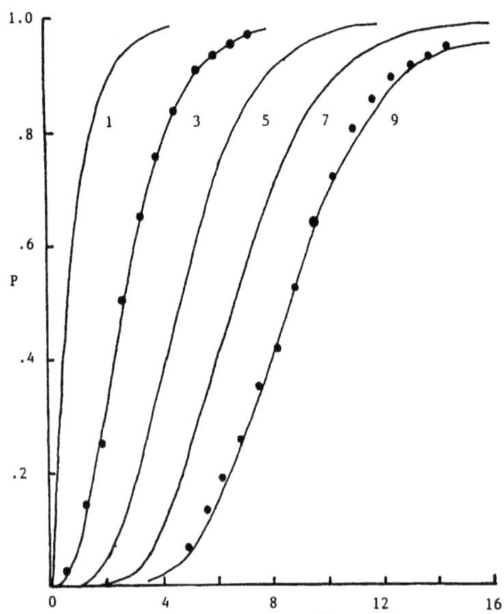

Fig. 8 Plot of $\sum_{k+1}^{\infty} P_j(T)$ versus T and comparison with experimental data.

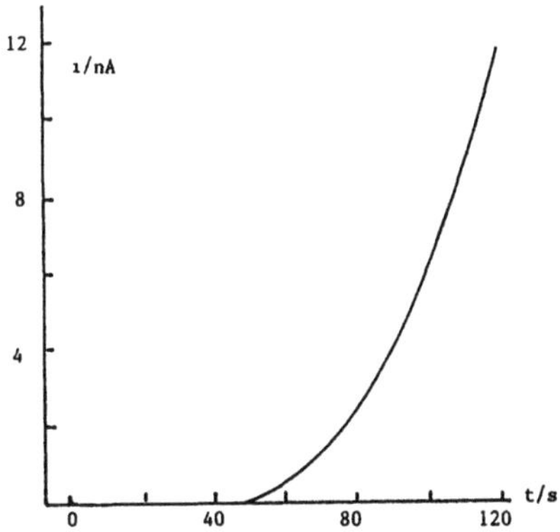

Fig. 9 Current-time transient for the deposition of a single centre of α-PbO_2 on 8μ diameter carbon disc microelectrodes: η = 400 mV; solution composition: 0.1M lead acetate + 1M sodium acetate + 1M acetic acid.

arrival time of the nucleus and, for this example of the oxidation of Pb^{2+} ions, n = 2. A comparison of the distribution of the arrival times with the predictions of the pure birth model is made for results at two overpotentials in fig. 8. It can be seen that there is excellent agreement for k = 3 at the highest overpotential whereas at lower overpotentials a larger cluster with k = 9 must be formed. At intermediate values of the overpotential the results lie between the two limits but the distributions are narrower than those predicted by the pure birth model. Since most simple modifications of the model predict a broadening of the distributions, it is necessary to assume that at these intermediate overpotentials a mixture of critical nuclei having different dimensions and structures is formed [20].

The lowest value of k observed (k = 3) can in fact be simply interpreted in terms of the columbite type structure of α-PbO_2. This consists of zig-zag chains of octahedra and the 'decision' to form α-PbO_2 requires the deposition of at least one zig-zag unit of the lattice, i.e. k = 3. Plausible explanations for k = 9 can also be put forward [20].

It can be seen therefore that the two-fold application of the concepts of measurements on very small systems (once to define the formation of a single growth centre and the second time to define the arrival time of a critical nucleus) allows the investigation of the kinetics of the initial stages of deposition at the molecular level. It should be noted that it is the subsequent essentially deterministic growth of the single centre (or indeed of a multiplicity of growth centres [22,23] which allows us to monitor the initial stochastic events i.e. the electrochemical processes themselves serve to amplify the signals due to the random processes. We note also that the mass transfer coefficient to the very small clusters becomes very large ($k_m \simeq$ 500 cm s^{-1}). In the limit this is the heterogeneous equivalent of the limiting diffusion controlled second order reaction rate constant ($k_2 \simeq 10^{10}$ litre mol^{-1} s^{-1}); in the case of a heterogeneous reaction the value can be calculated for collisions in a five or three-dimensional hyperspace (as compared to the six-dimensional hyperspace for an homogeneous reaction) depending on whether the cluster is mobile or stationary on the substrate surface. The very high value of this rate constant ensures that the initial stages of cluster formation will usually be under kinetic rather than diffusion control.

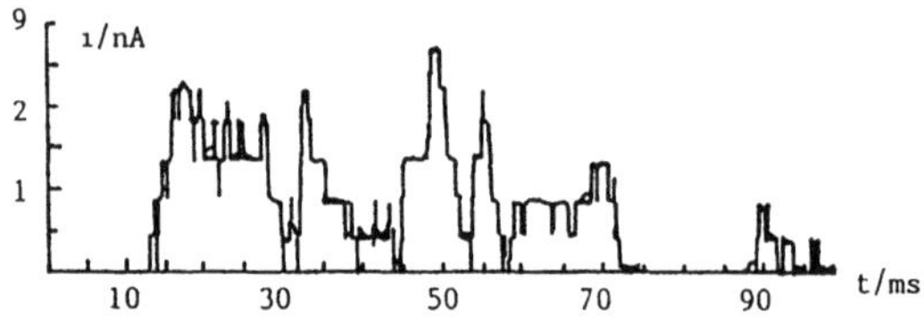

Fig. 10 Section of a current (i)-time (t) record. Bilayer formed from glycerol monooleate (0.005M) + cholesterol (0.01M) in hexadecane. Potential difference, 175 mV. Membrane area 0.4 cm^2. Temperature 294°. The respective synchronous reclassified sequences have been overlayed (using mean state conductances) distinguished by zero noise.

The concept of the amplification of the individual molecular events by subsequent essentially deterministic processes can be applied to numerous other phenomena and we illustrate the principle with a further example of nucleation drawn from the field of bioelectrochemistry. It is well known that the insertion of 'pore formers' into lipid bilayers leads to transmembrane ion currents which show discrete levels of conductance (for review see e.g. [24]). The changes, which in some cases are gated by the application of appropriate potential differences across the bilayers, are attributed either to reorientation or the insertion/aggregation/deaggregation of the pore former. Fig. 10 illustrates the current-time series observed for the polypeptide pore former alamethicin which can be attributed to the nucleation of pores in the membrane [25-28]: evidently the magnitude of the changes in the current due to the increase/decrease in dimensions of the pore are sufficiently large that the individual molecular events become measurable. The kinetics of the transitions between different states in the # space can be obtained directly from the transition probabilities while the interpretation of the duration of the occupation of particular levels in terms of a Boltzmann-type distribution gives the energetics of a particular state. Fig. 11 shows that the electrochemical free energy is initially a parabolic function of #n as would be predicted for the nucleation of a circular pore. However, the free energy is always positive and, moreover, the surface energy due to the replacement of lipid by solution is positive whereas the edge energy due to the insertion of the pore former is negative. The pore must therefore reach an upper limiting size i.e. an upper reflecting state.

As the kinetics of transition and the differences of free energy can both be directly determined, it is possible also to probe the relation between the two. The expected linear free energy relationship is observed [28] and, if we interpret the kinetics in terms of a relation such as equation (2) (with ηF replaced by $N\,\Delta\bar{g}$ where N is Avogadro's number) we

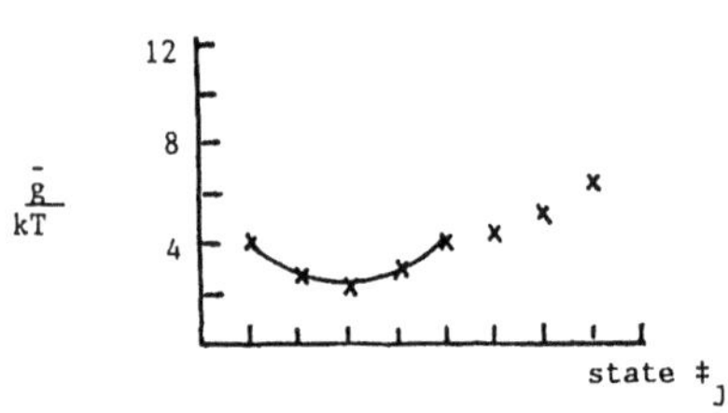

Fig. 11 Free energy vs state cardinal. Mono-olein system. The solid line marks a quadratic function fitted by least squares to the observed values (*).

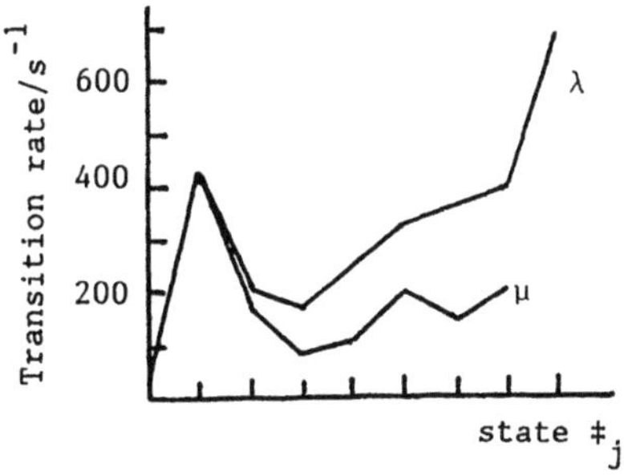

Fig. 12 Rate parameters λ_j and μ_j versus state. Mono-olein system.

Fig. 13 Model of pore expansion: insertion of a dimer (alamethicin/alamethicin or alamethicin/lipid).

obtain $\alpha \simeq 1$ for the birth rates λ_j. Surprisingly, however, the death rates vary in the same way with Δ_g as do the rates of birth, fig. 12, i.e. $(1-\alpha)$ #0. The only possible interpretation is that pore expansion/contraction takes place via an intermediate more highly expanded state, fig. 13.

It should be noted that measurements of this kind allow us to define an 'electrode' of molecular discussions even though the supporting structure (the bilayer) is comparatively large $\sim 1\ mm^2$. At higher levels of activity we will clearly form a multiplicity of pores so that we observe deviations from the simple relationships e.g. the behaviour in fig. 11 at high #n. However, it is possible to deconvolute the behaviour of such a compound system of a multiplicity of pores into that of the elementary system (the single pore which we model) and a trigger rate and such an analysis in fact shows that the upper reflecting state is at #4 or 5 [29,30]. Evidently, it should become possible to carry out comparable thermodynamic and kinetic analyses on other systems where the observed variables amplify individual atomic or molecular events (e.g. in field ion microscopy) so that detailed information about the initial stages of cluster formation should become more generally accessible.

4. The Growth of Crystallites

An example of the three-dimensional growth of a single centre has already been given above (cf. equation (14)). On electrodes of conventional size (say 1 cm^2), the time transient must be predicted by combining that for a single centre with an appropriate nucleation law. It is found that nucleation usually takes place at a number $N_o(cm^{-2})$ of preferred sites and to a first approximation can be described by

$$N = N_o\,(1 - \exp(-At)) \tag{15}$$

where A (s^{-1}) is the nucleation rate constant. In the initial stages of growth we can distinguish two limiting cases: the nucleation rate constant is either small or large compared to the time scale considered. In the first case

$$N \simeq AN_o t \qquad (16)$$

and in the second

$$N \simeq N_o \qquad (17)$$

Combination of either (16) or (17) with (14) shows that depending on which limit applies, we will observe a transient

$$i \simeq \frac{2\pi nFM^2k^3AN_o t^3}{3\rho^2} \qquad (18)$$

or

$$i \simeq \frac{2\pi nFM^2k^3N_o t^2}{\rho^2} \qquad (19)$$

Moreover, as the rate constant of nucleation varies very rapidly with the overpotential (cf. equation (9)) we can change the transient (18) to (19) by holding the electrode for a short time at a high overpotential before measuring the kinetics of crystal growth. Thus the rate constants A and k^3N_o can be independently determined and interpreted (e.g. see [19]). In the later stages of crystal growth it is usually necessary to consider the complete form (15) and to allow for the overlap of growth centres by Avrami's theorem [31] e.g. see [32,33].

Comparable expressions for the rate of growth of two-dimensional growth centres are readily written down. The most important cases concern the electrocrystallisation of a single layer plane of a crystal e.g. [17, 34-37] or the first order (or even higher order) phase transformation of adsorbates [24,38]. In most of these examples there are no preferred sites for nucleation so that we redefine the nucleation rate constant as A ($cm^{-2}\ s^{-1}$) and the most likely form for the initial stages of growth is

$$i \simeq \frac{\pi nFMhk^2At^2}{\rho} \qquad (20)$$

If the nucleation is confined to N_o centres

$$i \simeq \frac{2\pi nFMhk^2N_o t}{\rho} \qquad (21)$$

Allowing for the overlap of the centres, we have for the formation of a single layer plane the corresponding expression

$$i = \frac{\pi nFMhk^2At^2}{\rho} \exp\left(-\frac{\pi M^2k^2At^3}{3\rho^2}\right) \qquad (22)$$

and

$$i = \frac{2\pi nFMhk^2N_o t}{\rho} \exp\left(-\frac{\pi M^2kN_o t^2}{\rho^2}\right) \qquad (23)$$

For the simplest possible situation of the growth of individual two-dimensional centres on a microelectrode substrate, such as the deposition of silver on perfect (100) Ag single crystal faces we observe a time series of the current such as that in fig. 14 [34,39]. In this case the growth of

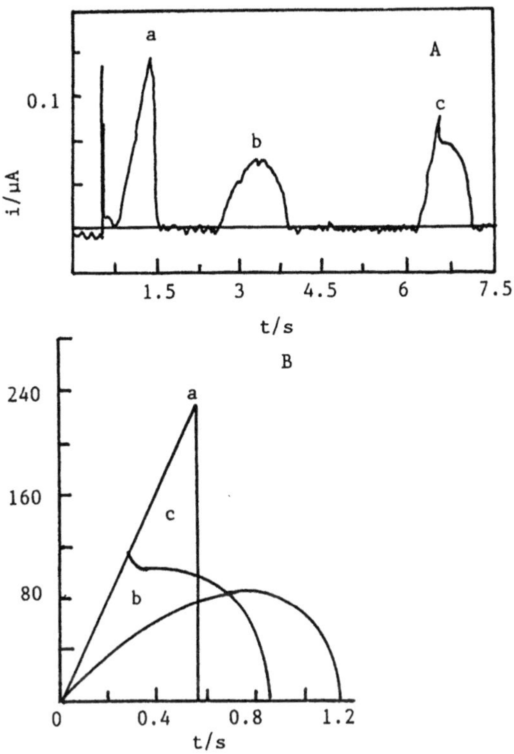

Fig. 14 (A) Section of data of the initial stages of deposition; overpotential 8 mV, nucleation at (a) the centre, (c) midway between the centre and the edge, (b) the edge. B. The predicted shapes of the current-time transients for the growth of circular centres on a circular substrate electrode. Nucleation (A) at the centre, (B) midway between the centre and the edge (c) the edge.

the single centre is not limited by overlap but by the propagating edges reaching the boundaries of the substrate which in this case has a circular shape. If the growth centre is itself circular (or highly polygonised) then the shapes of the transients a, b and c can be predicted in terms of the locations of the birth of the nuclei, see inset fig. 14. It should be noted that the shape and dimension of the small substrate area now become parameters of the investigation. It is of special interest to analyse the higher moments of the reaction rates alone (i.e. of the ensemble of transients)

$$\langle i^n(t)\rangle = \int_0^R i^n\,(r,b)P(b)db \qquad (24)$$

where P(b) is the probability density for the birth of a crystallite at a distance b (cm) from the centre, r (cm) is the radius of the centre at time t and R (cm) is the radius of the substrate electrode. The mean and second, third and fourth moments of the experimental data about the mean are compared in fig. 15 with the theoretically predicted values based on a

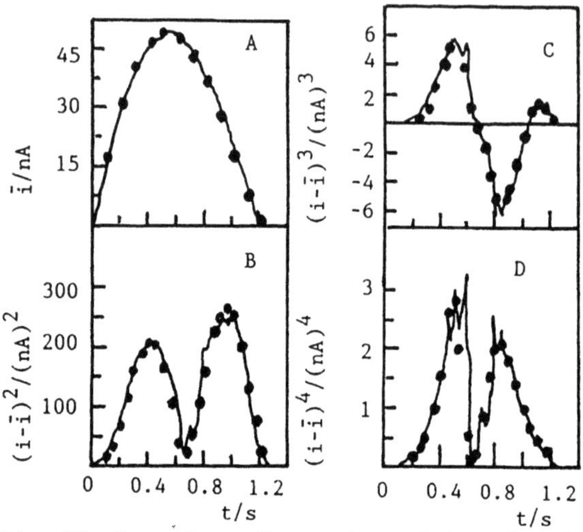

Fig. 15 Comparison of experimental data for the mean and the second, third and fourth moment about the mean current with the theoretically predicted moments for the model assuming a uniform probability of nucleation in space. (A) The mean, (B) the second moment about the mean, (C) the third moment about the mean, (D) the fourth moment about the mean. (——) experimental data. (.....) Predicted data.

model of circular growth centres randomly nucleated on the surface and with the slow step of the reaction being at the edges of the growth centre. In contrast to the first moment (the mean) the higher moments of the reaction rate are highly sensitive to the nature of the assumptions (note for example the complex behaviour of the third moment about the mean) so that the agreement between the experimental and calculated results fully substantiates the model: in particular any retardation due to slow surface or bulk diffusion of the reactant Ag^+ species is excluded.

The predictions of the transients for multinuclear growth equations (22) and (23), are based on the assumption of an effectively infinite substrate. It is in fact possible to extend the Avrami formalism to allow the calculation of the shapes of the transients when both the dimensions and shape of the substrate electrode and overlap are significant in restricting the growth of centres and measurements under such conditions significantly extend the scope of the kinetic investigations [40]. At the present time, however, most investigations have been carried out under conditions where the assumption of an infinite substrate are effectively valid and equations (22) and (23) have been shown to apply to a wide variety of systems e.g. [35,38,41] including the anodic deposition of salts and oxides e.g. [17, 35-37] on perfectly smooth mercury or amalgam substrates, underpotential deposition of metals [38] on smooth single crystal substrates and phase transformations in adsorbed layers [24]. Fig. 16 compares the experimenttal data with the prediction (22) for one such system, the formation of a monolayer of HgS [37] while fig. 17 illustrates a completely different type of system, the oxidation of CO adsorbed on Pt [41] which evidently proceeds by the nucleation of 'holes' in a coherent layer. In this case it would be expected that nucleation is initiated at preferred sites and the complete set of data in fact fit the transient predicted for the combina-

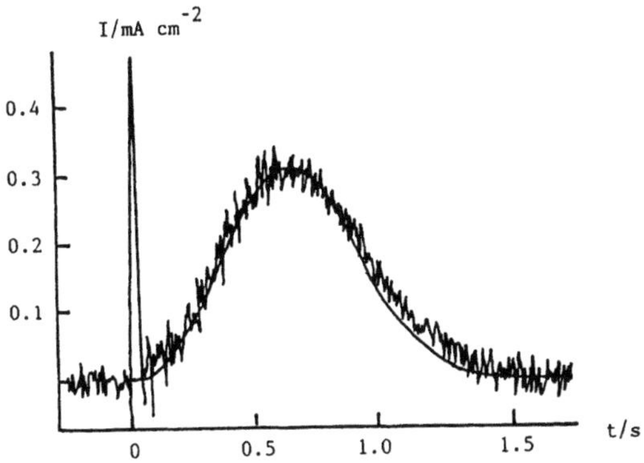

Fig. 16 Experimental transient for the deposition of a monolayer of HgS on a mercury electrode in 0.1 mol dm^{-3} Na_2S - 1.0 mol dm^{-3} $NaHCO_3$, illustrating the fit predicted by Equation (22) for the progressive nucleation of a monolayer. The charge under the transient corresponds to the formation of a monolayer of HgS in the (001) plane.

tion of equation (15) with a model of the growth of holes which overlap according to the Avrami theorem:

$$i = \frac{2\pi nFMhk^2N_o}{\rho}\left\{t - \frac{1}{A}\left[1 - \exp(-At)\right]\right\}$$

$$\times \exp\left\{-\frac{\pi M^2k^2N_o}{\rho^2}\left[t^2 - \frac{2t}{A} + \frac{2}{A^2}\left(1 - \exp(-At)\right)\right]\right\} \quad (25)$$

It will be evident that phase growth frequently takes place in these systems in highly defined crystallographic directions and this can readily be confirmed using replica techniques and selected area electron microscopy and diffraction. Fig. 18 illustrates an example of such measurements, the successive deposition of two monolayers of TlCl on thallium amalgam, followed by the deposition of a multi-molecular layer (thickness equivalent to $\sim$ 10 monolayers [17]. Deposits formed at low overpotentials (< 80 mV) show the absence of the diffractions associated with the (111) planes of TlCl, fig. 18A. This is clear evidence that the deposit is orientated with the (100) planes parallel to the substrate and is confirmed

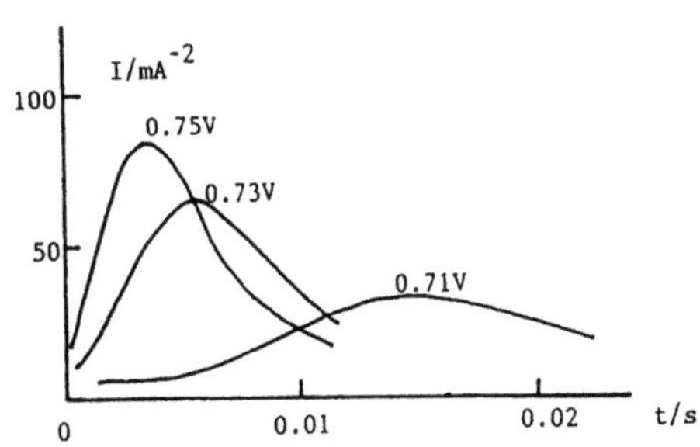

Fig. 17 Oxidation of CO adsorbed onto a Pt electrode from an 1.0M $HClO_4$ solution saturated with CO. Potential steps to values shown (versus S.C.E.)

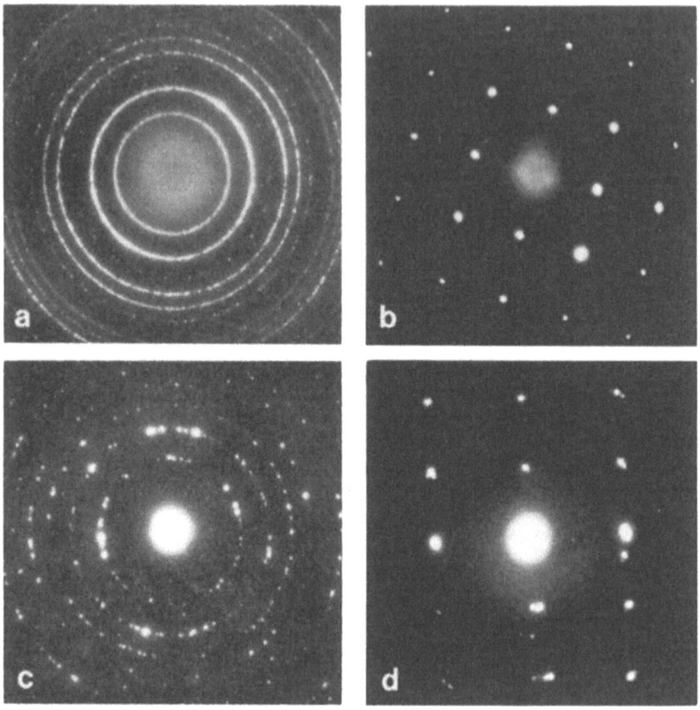

Fig. 18 Electron diffraction patterns of TlCl formed on 1% Tl/Hg in 1M HCl. (A) at η = 40mV for 100 ms; (B) at η = 50 mV for 30 ms; (C) at η = 120 mV for 10 ms; (D) at η = 90 mV for 10 ms. (A) and (B) show [100] orientation, (D) shows [110] orientation.

by the occasional observation from larger single crystal areas, fig. 18B. Deposits prepared at η > 80 mV show diffractions from all crystal planes, fig. 18C, but diffractions from essentially single crystal areas show conclusively that much of the deposit is orientated with the prism face (110) parallel to the substrate, fig. 18D.

In a few cases, such as the deposition of monolayers of Ni on perfectly smooth mercury substrates, fig 19, the current-time transients show characteristic distortions (such as the observation of current maxima at low coverages) which can be attributed to the overlap of diffusion zones surrounding the growth centres rather than to the overlap of the growth centres themselves, fig. 20, [16]. Changes of shape with the applied potential can be interpreted in terms of alterations in the width of the diffusion zones caused by the changes in the mean lifetime of the adatoms.

5. Measurements on Electrode-Solution Interfaces at the Molecular Level

Although, as has been shown, it is possible to develop electrochemical techniques to the point where events can be resolved at the molecular level, these techniques by themselves lack molecular specificity. Such specificity can only be achieved by combining the kinetic measurements

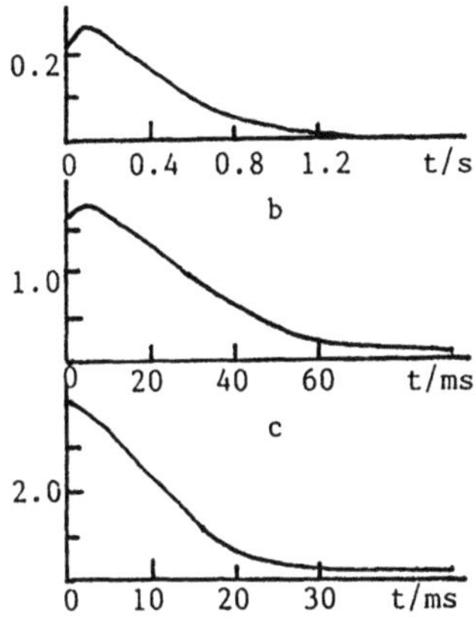

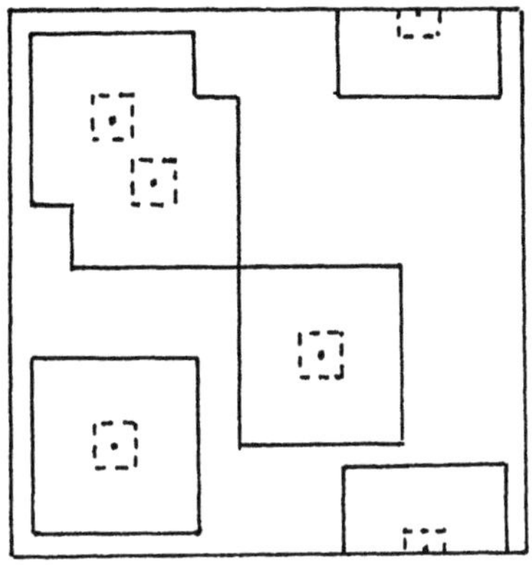

Fig. 19 Typical cathodic i-t transients for deposition of a monolayer of Ni onto a Hg electrode from a 0.1M $NiSO_4$ + 0.1M KSCN + 2.2M KNO_3 solution. Area of surface 0.055 cm^2. Potentials: (a) -600 mV, (b) -640 (mV), (c) -670 mV versus S.C.E.

Fig. 20 Typical simulation diagram for instantaneous nucleation of 6 centres. The dashed lines enclose the growing centres; the areas between the full and dashed lines contain the diffusion zones.

with appropriate spectroscopic and diffraction probes. The example of fig. 18 discussed above is in fact an early application of what has become known as the 'ex-situ' examination of electrode-solution interfaces (however, it should be noted that RHEED measurements on electrodeposits were being developed extensively already before the second world war). More recently ex-situ investigations have relied on the highly sensitive electron spectroscopies, diffraction and imaging techniques, which are coupled to the electrochemical system by the use of appropriate transfer techniques (for reviews see e.g. [42-44]). Such measurements are naturally restricted to the properties of the electrode surfaces and of strongly bound adsorbates. In situ investigations, which are being developed at the same time can, in addition monitor the behaviour of weakly bound species and of the solution side of the interfacial regions. However, such measurements have to rely on the absorption, scattering and, in some instances, diffraction of photons and of a limited number of fundamental particles. The information sought includes the identification ('finger printing') of the species present (by no means an easy task), the determination of the energetics of these species (electronic, vibrational and rotational levels) and the determination of short range and long range order. Such information is sought both for systems in equilibrium and, increasingly for reactants, intermediates and products to aid the interpretation of reaction mechanisms. Vibrational spectroscopy is evidently well suited to the determination of such molecular properties since the spectra both identify the species present and are sensitive to the nature of the environment (bonding, interactions with other species: molecules as 'tell-tales'). The range of systems studied by infra-red spectroscopy is covered in a number of reviews e.g. [45,46] as is the application of Raman spectroscopy e.g. [47-50]. Here we restrict attention to a single example [51] drawn from the latter field to illustrate the scope of the method. Vibrational spectra, however, only give information about local order and orientation indirectly so that long range ordering must be investigated using diffraction techniques. Here we restrict attention to two examples [52] of the application of in-situ X-ray diffraction [53] to illustrate the scope of these measurements.

At first sight it would appear to be difficult, if not impossible to use Raman spectroscopy to investigate electrode-solution interfaces in view of the low cross-section for normal Raman scattering. However, high quality spectra of adsorbates such as pyridine have been shown to be observable [54] on specially roughened silver electrodes [55,56] and on the surfaces of metal overlayers on such roughened substrates [57-59]. The intensities are enhanced by one or more than one enhancement mechanism (Surface Enhanced Raman Scattering, SERS, see reviews in [47-50]) so that such measurements are surface selective. Figs. 21 and 22 illustrate one such system, the adsorption of the pseudo halide CN^- group (isoelectronic with CO) on silver which has been extensively investigated (e.g. [60-65] and references cited therein). The complexity of the spectra at low frequencies shows that at least two CN^- groups are adsorbed at one site, fig. 21. The pseudo halide group interacts strongly with water, fig. 22 and both the C≡N and O-H frequencies depend markedly on the applied potential. These shifts have been attributed to the pronounced back donation of electrons from the d orbitals of the metal to the π^* orbitals on the CN^- group and electrostatic interaction of water with the negative charges. Fig. 22 also shows that the intensities of the δ OH band at $\sim$ 1600 cm^{-1} become pronounced over a narrow range of potential, possibly due to clustering of the solvent in this region. Other structural transformations can be observed. Thus fig. 22 and 23 show that for the adsorption of $Ag(CN)_2^-$ ions there is a change in orientation with increasing negative potentials before the start of electrodeposition, probably from an essentially C_{2v} to $D_{\infty h}$ configuration, fig. 24. This change in structure is not observed by normal electrochemical techniques such as the measurement of the differential capacitance. The structure of the adsorbed $Ag(CN)_2^-$ ion resembles that of adsorbed CN^- at more negative potentials except that $Ag(CN)_2^-$ does not interact with water. It is evident that SERS can now be used to follow the reaction paths of electrochemical processes.

The determination of the structure of interfacial regions by transmission and reflection X-ray diffraction [52,53] would appear to be even more difficullt than the use of Raman spectroscopy. However, the

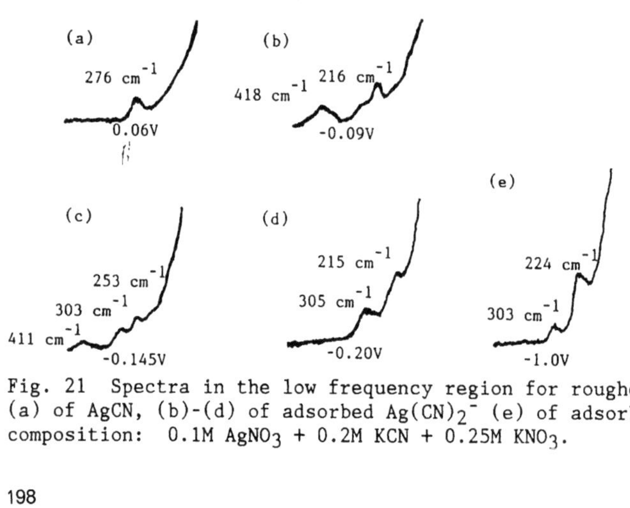

Fig. 21 Spectra in the low frequency region for roughened silver surfaces, (a) of AgCN, (b)-(d) of adsorbed $Ag(CN)_2^-$ (e) of adsorbed CN^-. Solution composition: 0.1M $AgNO_3$ + 0.2M KCN + 0.25M KNO_3.

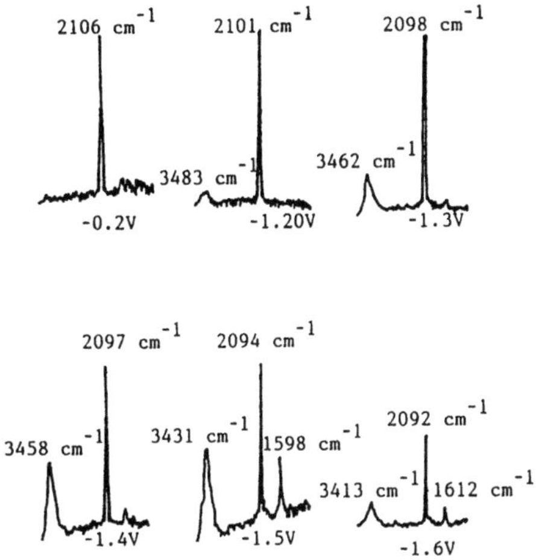

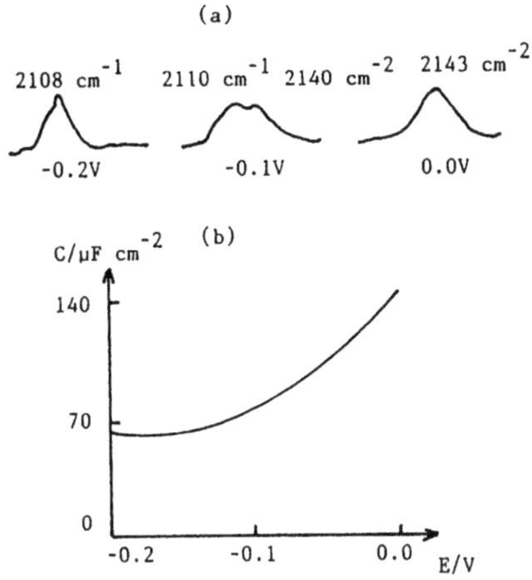

Fig. 22 Spectra in the (OH) region for roughened silver surface: (a) adsorbed $Ag(CN)_2^-$, (b)-(e) adsorbed CN^-. Spectra measured at constant potential using a Coderg T800 spectrometer.

Fig. 23 (a) Raman spectra of the C-N stretch region from a solution of 0.025M $AgNO_3$, 0.050M KCN and 0.50M KNO_3 at the potentials indicated in the figure and (b) corresponding differential capacitance curve in the same potential region. Data determined using a linear potential sweep at 5 mV s^{-1}; spectra measured with Anaspec 36 spectrograph at the potentials shown during a linear sweep and with 1s acquisition time.

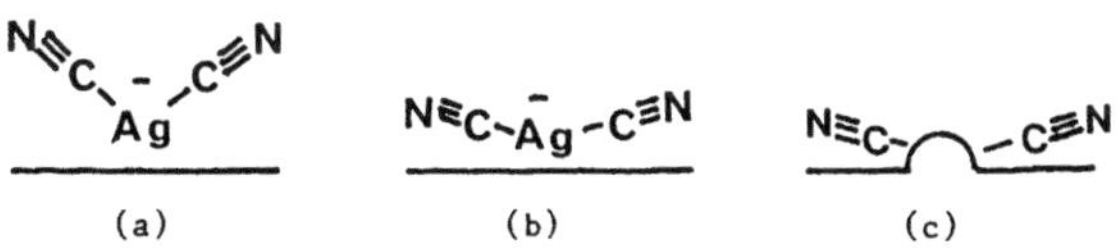

Fig. 24 (a) and (b) Model for the adsorption of $Ag(CN)_2^-$. (c) model for adsorbed CN

development of position sensitive single photon count systems [66] and the access to the required on-line computing now makes this seemingly impossible task feasible. Surface sensitivity in electrochemical systems can be enhanced when necessary by applying the modulation and signal averaging techniques which have already been extensively established for in-situ spectroscopy (see Introduction). Fig. 25 illustrates the experimental arrangement for reflection diffraction. Here we illustrate measurements of changes induced by adsorption at Pt-sulphuric acid interfaces (cf. fig. 3). Fig. 26A shows that there are no detectable changes in structure during the formation of 'strongly adsorbed' hydrogen at least at the level of sensitivity which can be achieved with the current instrumentation. On the other hand fig. 26B shows that there are marked changes in structure for the formation of 'weakly adsorbed' hydrogen. The two main diffractions of bulk Pt(111) and (200) are differentiated due to simultaneous loss of intensity for the diffraction from the bulk of the crystallites and the appearance of new diffraction lines due to ordered but incommensurate surface arrays. For a surface such as that of Pt consisting only of atoms

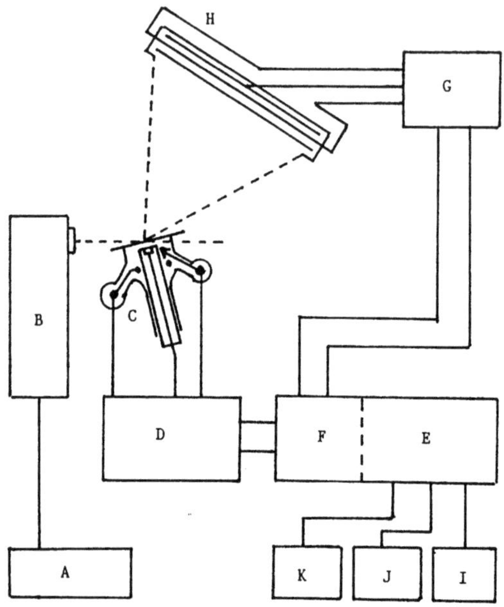

Fig. 25 Schematic diagram of reflection X-ray diffraction apparatus. A - High voltage supply; B - X-ray tube; C - Electrochemical cell; D - Potentiostat; E - Microcomputer; F - CAMAC crate; G - NIM bin; H -Detector; I - Disc drives; J - VDU monitor; K - Printer.

of a single species, we can only observe diffractions for primitive Bravais nets, i.e. a very simple pattern due to low index diffractions. The line at $\sim 38.8^{\circ}$ is readily explained as the (11) diffraction of a close packed surface layer, the Pt-Pt distance having the expected value 0.274 nm. However, the line at $\sim 46.0^{\circ}$ corresponds to such a short spacing that it is necessary to assume an expansion of the surface layer to permit the observation of such a shortened (10) row spacing. In fig. 27 we suggest a possible model. Repeated modulation of the electrode potential is known to lead to the development of (110) faces [69] and the structure in fig. 27 could be produced by a surface relaxation of slightly expanded (110) faces, involving also the next underlying layer [68]. The formation of such incommensurate surface layers will naturally lead to the loss of diffraction intensity due to the bulk diffractions; we observe that the changes of structure are reversible. We also note that deconvolution of the difference diffractograms gives diffraction lines due to the surface species which have half widths which are comparable to that of the bulk diffractions (corresponding to a coherence length of $\sim$ 15 nm). It is evident therefore that these surface structures are formed on each crystallite but that two types of surface lattice are simultaneously present.

Changes in structure can also be observed for the adsorption of other adsorbates such as CO. Fig. 28 shows that, as for the case of the adsorption of weakly bound hydrogen, the (111) and (200) diffractions are differentiated. As in that case the line at $\sim 38.8^{\circ}$ corresponds to the (11) diffraction of a close packed layer while that at $\sim 46.0^{\circ}$ corresponds to the (10) diffraction of an expanded lattice and it may well be that this

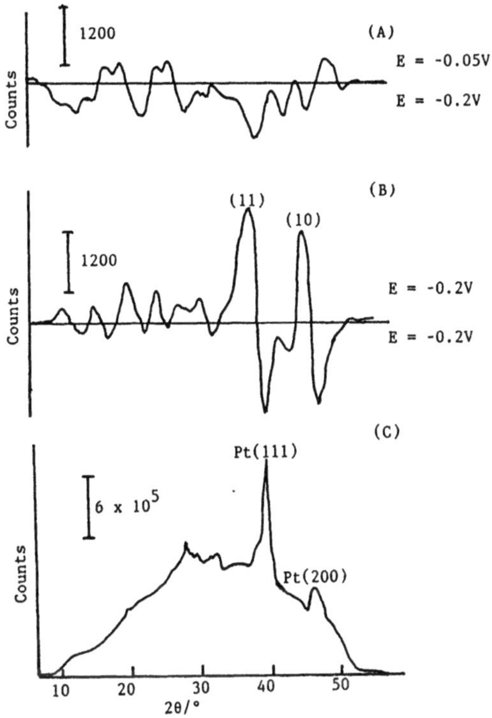

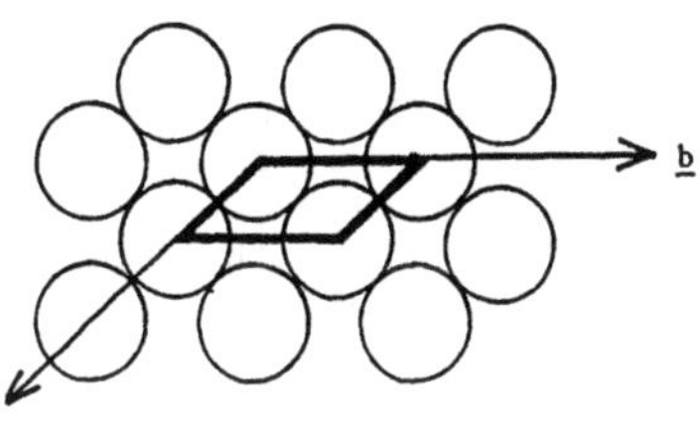

Fig. 26 (C) Reflection X-ray diffractogram from platinised Pt in 1M H_2SO_4 in double layer region. (A) and (B) Reflection difference diffractograms from platinised Pt in 1M H_2SO_4: (A) strongly adsorbed hydrogen region; (B) weakly adsorbed hydrogen region. Modulation at 0.004 Hz between potentials shown; data collection time 3×10^4s.

Fig. 27 Model for one of the structures of the platinised Pt surface layers formed by the adsorption of weakly adsorbed hydrogen. The angle between the vectors $\underline{a}$ and $\underline{b}$ of the unit cell is $\sim 134.0^{\circ}$.

type of behaviour will be found to be common to a wider range of adsorbates. It should be noted that the diffractograms given in fig. 28 correspond respectively to the adsorption conditions giving respectively CO coordinated to single Pt atoms and that coordinated to several atoms. The transition to the second species is associated with a further expansion of the parts of the surface giving rise to the (10) diffraction. However, both types of surface species clearly cause comparable types of structural changes which again are reversible. These structural changes (especially the facile formation of expanded surface layers) may well be responsible for the high catalytic activity of Pt-surfaces for a wide range of heterogeneous reactions.

Closure Most practical electrochemical systems consist of ensembles of microelectrodes e.g. the facets of a microcrystalline platinised platinum electrode or the crystallites of a growing electrodeposit. It is now possible to study electrode reactions on individual microelectrodes and, in suitable cases, such measurements give information about kinetics at the

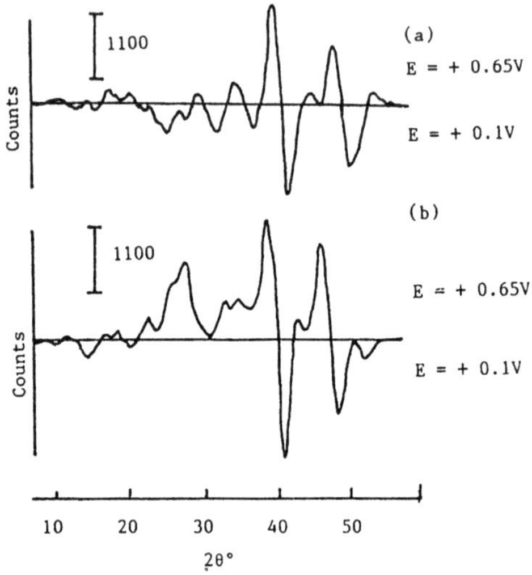

Fig. 28 Reflection difference X-ray diffractograms from platinized Pt in 0.1M H_2SO_4 saturated with CO. (a) Potential modulated between +0.1V and +0.65V. Data collection time 3 x 10^4s. (b) Potential modulated between +0.1V and 0.65V with longer waiting time (150s) at +0.1V. Data collection time 2.5 x 10^4s.

molecular level. The interpretation of this information usually requires a detailed structural model. The in-situ investigation of the interfacial regions by vibrational spectroscopic and diffraction methods gives information about the nature, energetics and structure of the species present and this detailed information can be obtained for practical working systems i.e. ensembles of microelectrodes. In suitable cases the in-situ techniques in turn give information about reaction mechanisms.

References

1. D.C. Grahame, Chem. Rev., 41, 441 (1947).
2. A.N. Frumkin, P.I. Dolin and B.V. Ershler, Acta Physicochim. U.R.S.S. 13, 779 (1940).
3. F.B. Bowden and E.K. Rideal, Proc. Roy. Soc., A120, 59 (1928).
4. G. Armstrong and J.A.V. Butler, Trans. Faraday Soc., 29, 1261 (1933).
5. A.N. Frumkin and A. Slygin, C.R. Acad. U.R.S.S, 2, 173 (1934).
6. P. Dolin and B. Ershler, Acta Physicochim. U.R.S.S., 13, 747 (1940).
7. Ultramicroelectrodes Eds. M. Fleischmann, S. Pons, D.R. Rollison and P.P Schmidt, Datatech Systems Inc., P.O. Box 435, Morgantown, N.C., U.S.A. (1987).
8. K.B. Oldham, Trans. Faraday Soc., 53, 80 (1957).
9. G.C. Barker in E. Yeager (Ed.) Transactions of the Symposium on Electrode Processes, Philadelphia (1959), J. Wiley, New York, 325 (1961).
10. M. Fleischmann and S. Pons, reference [7], Chapter 2.
11. A.M. Bond, M. Fleischmann, S.B. Khoo, S. Pons and J. Robinson, Indian J. of Technology, 24 492 (1986).
12. P. Bindra, A.P. Brown, M. Fleischmann and D. Pletcher. J. Electroanal. Chem., 58, 31 (1975).

13. P. Bindra, A.P. Brown, M. Fleischmann and D. Pletcher, J. Electroanal. Chem., 58, 39 (1975).
14. G. Gunawardena, G.J. Hills and B. Scharifker, J. Electroanal. Chem., 130, 99 (1981).
15. S. Toschev and I. Markov, Electrochim. Acta, 12, 281 (1967).
16. M. Fleischmann, J.A. Harrison and H.R. Thirsk, Trans. Faraday Soc., 61, 2742 (1965).
17. M. Fleischmann, J. Pattison and H.R. Thirsk, Trans. Faraday Soc., 61, 1256 (1965).
18. M. Fleischmann and H.R. Thirsk in Advances in Electrochemistry and Electrochemical Engineering, Eds. P. Delahay and C.W. Tobias, Vol. 3, 123 (1963). Interscience - John Wiley, New York and London.
19. J.B. Zeldovich, Acta Physicochim. U.R.S.S. 18, 1 (1943.
20. M. Fleischmann, L.J. Li and L. Peter, to be published.
21. M. Fleischmann, L.J. Li and L. Peter, 15th International Power Sources Symposium, Brighton, England, September (1985), Power Sources Ed. L.J. Pearce, Vol. 11 (1987).
22. M. Fleischmann, M. Labram, C. Gabrielli and A. Sattar,Surf. Sci., 101, 583 (1980).
23. M. Fleischmann, C. Gabrielli and M. Labram, J. Electroanal. Chem., 150 111 (1983).
24. R. de Levie in Advances in Electrochemistry and Electrochemical Engineering Eds. H. Gerischer and C.W. Tobias, Vol. 13, 1 (1984), Interscience - John Wiley, New York and London.
25. L.G.M. Gordon and D.A. Haydon, Biochim. Biophys. Acta, 255, 1014 (1972).
26. M. Eisenberg, J.E. Hall and C.A. Mead, J. Membrane Biol., 14, 143 (1973).
27. P. Mueller and D.O. Rudin, Nature, 217, 713 (1967).
28. M. Fleischmann, C. Gabrielli, M.T.G. Labram, A.I. McMullen and T.H. Wilmshurst, J. Membrane Biol. 55, 9 (1980).
29. M. Fleischmann, C. Gabrielli, M.T.G. Labram and T. Markvart, J. Electroanal. Chem., 214, 427 (1986).
30. M. Fleischmann, C. Gabrielli and M.T.G. Labram, J. Electroanal. Chem., 214, 441 (1986).
31. M. Avrami, J. Chem. Phys., 7, 1103 (1939); 8, 212 (1940); 9, 177 (1941).
32. R.D. Armstrong, M. Fleischmann and H.R. Thirsk, J. Electroanal. Chem., 11, 208 (1966).
33. M.Y. Abyaneh and M. Fleischmann, J. Electroanal. Chem., 119, 187 (1981); 119, 197 (1981).
34. E. Budevski, W. Bostanoff, T. Witanoff, Z. Stoinoff, Z. Kotzewa and R. Kaischev, Electrochim. Acta, 11, 1697 (1966).
35. A. Bewick, M. Fleischmann and H.R. Thirsk, Trans. Faraday Soc., 58, 2200, (1962).
36. M. Fleischmann, K.S. Rajagopalan and H.R. Thirsk, Trans. Faraday Soc., 59, 741 (1963.
37. L.M. Peter, J.D. Reid and B.R. Scharifker, J. Electroanal. Chem., 119, 73 (1981).
38. A. Bewick and B. Thomas, J. Electroanal. Chem., 70, 329 (1976); 84, 127 (1977); 85, 329 (1977).
39. E. Budevski, M. Fleischmann, C. Gabrielli and M.T.G. Labram, Electrochim. Acta, 28, 925 (1983).
40. M.Y. Abyaneh and M. Fleischmann, to be published.
41. C. McCallum and D. Pletcher, J. Electroanal. Chem., 70, 277 (1976).
42. A.T. Hubbard, Acc. Chem. Res., 13, 177 (1980).
43. P.M.A. Sherwood, Chem. Soc. Reviews, 1 (1985).

44. P.N. Ross and F.T. Wagner in Advances in Electrochemistry and Electrochemical Engineering, Eds. H. Gerischer and C.W. Tobias, Vol. 9, 101 (1984), Interscience - John Wiley, New York and London.
45. A. Bewick and S. Pons in Advances in Infra Red and Raman Spectroscopy Eds. R.J.H. Clark and R.E. Hester, 12, 1 (1985).
46. J. Foley, C. Korzeniewski, J. Daschbach and S. Pons in Electroanalytical Chemistry, Ed. A. Bard, 14, 309 (1986).
47. T.E. Furtak and R.K. Chang, Surface Enhanced Raman Scattering, Plenum Press, New York and London (1982).
48. M. Fleischmann and I.R. Hill in Comprehensive Treatise of Electrochemistry, Eds. R.E. White, J. O'M Bockris, B.E. Conway and E. Yeager, 12 (1984), Plenum Press, New York and London.
49. S. Efrima in Modern Aspects of Electrochemistry, Eds. J. O'M Bockris, R.E. White and B.E. Conway, 16, 253 (1985).
50. M. Moskovits, Reviews of Modern Physics, 57, 783 (1985).
51. M. Fleischmann, G. Sundholm and Z.Q. Tian, Electrochim. Acta, 31, 907, (1986).
52. M. Fleischmann and B.W. Mao, J. Electroanal. Chem., 229, 125 (1987).
53. M. Fleischmann, A. Oliver and J. Robinson, Electrochim. Acta., 31, 899 (1986).
54. M. Fleischmann, P.J. Hendra and A.J. McQuillan, Chem. Phys. Lett., 26, 163 (1974).
55. D.L. Jeanmaire and R.P. Van Duyne, J. Electroanal. Chem., 84, 1 (1977).
56. M.G. Albrecht and J.A. Creighton, J. Amer. Chem. Soc., 99, 5215 (1977).
57. M. Fleischmann, Z.Q. Tian and L.J. Li, J. Electroanal. Chem., 217, 393 (1987).
58. M. Fleischmann and Z.Q. Tian, J. Electroanal. Chem., 217, 411 (1987).
59. M. Fleischmann and Z.Q. Tian, J. Electroanal. Chem., 217, 385 (1987).
60. J. Billman, G. Kovacs and A. Otto, Surf. Sci., 92, 153 (1980).
61. R.E. Benner, R. Dornhaus, R.K. Chang and B.L. Laube, Surf. Sci., 101, 341 (1980).
62. R. Kotz and E. Yeager, J. Electroanal. Chem., 123, 335 (1981).
63. M. Fleischmann, I.R. Hill and M. Pemble, J. Electroanal. Chem., 136, 361 (1982).
64. T. Billman and A. Otto, Surf. Sci., 138, 1 (1984).
65. T.K. Chen, K.V. von Raben, D.V. Murphy, R.K. Chang and B.L. Laube, Surf. Sci., 143, 369 (1984).
66. C.J. Borkowski and M.K. Kopp, Rev. Sci. Inst., 46, 951 (1975).
67. K. Yamamoto, D.M. Kolb, R. Kotz and G. Lehmpfuhl, J.Electroanal. Chem., 96, 233 (1979).
68. M. Fleischmann and B.W. Mao, J. Electroanal. Chem., in the press.

Index of Contributors

Behm, R.J. 92
Blakely, J.M. 102
Block, J.H. 67
Blum, L. 172
Borgmann, D. 121

Chambers, G.P. 111
Cohen, P.I. 19

Doll, J.D. 144

Fine, J. 111
Fink, H.-W. 75
Fleischmann, M. 182
Freeman, D.L. 144

George, S.M. 2

Gordon II, J.G. 172
Grunze, M. 132

Hofmann, B. 121

Karimi, M. 43
Kreuzer, H.J. 63

Lagally, M.G. 53

Mackrodt, W.C. 155
Marton, D. 111
Melroy, O.R. 172

Petrich, G.S. 19
Pukite, P.R. 19

Schmidt, L.D. 37
Seebauer, E.G. 37

Tasker, P.W. 155
Tringides, M.C. 53

Umbach, C.C. 102
Unertl, W.N. 132

Vidali, G. 43
Voter, A.F. 144

Wedler, G. 121
Weimer, J.J. 132
Whaley, G.J. 19
Wu, P.K. 53

Zeitler, E. 72

MIX
Papier aus verantwortungsvollen Quellen
Paper from responsible sources
FSC® C105338

If you have any concerns about our products,
you can contact us on
ProductSafety@springernature.com

In case Publisher is established outside the EU,
the EU authorized representative is:
Springer Nature Customer Service Center GmbH
Europaplatz 3, 69115 Heidelberg, Germany

Printed by Libri Plureos GmbH
in Hamburg, Germany